Developing Sustainable Agriculture and Community

This book illustrates the ways in which communities can strengthen the links and set the stage for long-term partnerships between sustainable agriculture and sustainable rural community development initiatives. It provides lessons learned, first, from the community development literature that can help shape sustainable agriculture strategies, and second, from the sustainable agriculture literature that can prove useful in moulding sound and effective community development strategies.

The threads that weave the chapters together is the commitment to building and expanding the community capital resources that have important bearing on the sustainability of agriculture and the broader community of which it is a part. Certainly, the success of the agriculture/community partnerships is rooted in one critical ingredient – "social capital." To be effective over the long-term, sustainable development depends on a network of people, drawn from a wide array of interests, who have a strong trusting relationship with one another, and who are willing to work together in responding to the economic, environmental, and social challenges facing agriculture and community alike. At the same time, strategies that work to strengthen the stock of all seven types of community capitals are important to pursue. It is balanced investments in all seven types of community capitals that will contribute to the emergence of "community agency" – the ability of local people to act in a proactive manner in managing, utilizing, and enhancing local resources. With the emergence of "community agency," an important step in the pursuit of a sustainable future for both agriculture and community is possible.

This book was published as a special issue of *Community Development*.

Lionel J. Beaulieu is Director of the Purdue Center for Regional Development and Professor in the Department of Agricultural Economics, Purdue University.

Jeffrey L. Jordan is a Professor in the Department of Agricultural and Applied Economics at the University of Georgia and Director of the Southern Region USDA's Sustainable Agriculture Research and Education (S-SARE) Program.

Developing Sustainable Agriculture and Community

Edited by
Lionel J. Beaulieu and Jeffrey L. Jordan

Routledge
Taylor & Francis Group
LONDON AND NEW YORK

COMMUNITY DEVELOPMENT
Journal of the
Community Development Society

First published 2013
by Routledge
4 Park Square, Milton Park, Abingdon, Oxon OX14 4RN
605 Third Avenue, New York, NY 10017

Routledge is an imprint of the Taylor & Francis Group, an informa business

This book is a reproduction of *Community Development,* vol. 38, issue 3. The Publisher requests to those authors who may be citing this book to state, also, the bibliographical details of the special issue on which the book was based.

British Library Cataloguing in Publication Data
A catalogue record for this book is available from the British Library

ISBN 13: 978-0-415-63371-0 (hbk)

Typeset in Times New Roman
by Taylor & Francis Books

Publisher's Note
The publisher would like to make readers aware that the chapters in this book may be referred to as articles as they are identical to the articles published in the special issue. The publisher accepts responsibility for any inconsistencies that may have arisen in the course of preparing this volume for print.

Contents

Citation Information vii

1. Introduction: Sustainable Development: Connecting Agriculture and Community
Lionel J. Beaulieu and Jeffrey L. Jordan 1

2. Small Farm Clusters and Pathways to Rural Community Sustainability
Kathryn J. Brasier, Stephan Goetz, Lindsay A. Smith, Molly Ames, Joanna Green, Tim Kelsey, Anu Rangarajan, and Walt Whitmer 8

3. Connecting Sustainable Agriculture to Rural Development: The Case of Pasture-Based Dairy Grazing
Lois Wright Morton and Leah Miller 23

4. Food System Makers: Motivational Frames for Catalyzing Agri-Food Development through Multi-Stakeholder Collaboration
Wynne Wright, Michael Score, and David S. Conner 39

5. Linking Small Farms to Rural Communities with Local Food: A Case Study of the Local Food Project in Fairbury, Illinois
S.A. Hultine, L.R. Cooperband, M.P. Curry and S. Gasteyer 61

6. Linking Sustainable Agriculture and Community Development: The Lowcountry Food Bank's Use of Locally Grown Foods
Kenneth L. Robinson, Kathleen K. Robinson, Carlos Carpio, and David Hughes 77

7. Options for the Economic Health of Farmers, Farmers Markets and Communities: Homebased Fruit and Vegetable Microprocessing
Sandra Bastin 91

Index 101

Citation Information

The chapters in this book were originally published in *Community Development*, volume 38, issue 3 (Autumn 2007). When citing this material, please use the original page numbering for each article, as follows:

Chapter 2
Small Farm Clusters and Pathways to Rural Community Sustainability
Kathryn J. Brasier, Stephan Goetz, Lindsay A. Smith, Molly Ames, Joanna Green, Tim Kelsey, Anu Rangarajan and Walt Whitmer
Community Development, volume 38, issue 3 (Autumn 2007) pp. 8-22

Chapter 3
Connecting Sustainable Agriculture to Rural Development: The Case of Pasture-Based Dairy Grazing
Lois Wright Morton and Leah Miller
Community Development, volume 38, issue 3 (Autumn 2007) pp. 23-38

Chapter 4
Food System Makers: Motivational Frames for Catalyzing Agri-Food Development through Multi-Stakeholder Collaboration
Wynne Wright, Michael Score and *David S. Conner*
Community Development, volume 38, issue 3 (Autumn 2007) pp. 39-60

Chapter 5
Linking Small Farms to Rural Communities with Local Food: A Case Study of the Local Food Project in Fairbury, Illinois
S.A. Hultine, L.R. Cooperband, M.P. Curry and *S. Gasteyer*
Community Development, volume 38, issue 3 (Autumn 2007) pp. 61-76

Chapter 6
Linking Sustainable Agriculture and Community Development: The Lowcountry Food Bank's Use of Locally Grown Foods
Jan Vandemoortele Kenneth L. Robinson, Kathleen K. Robinson, Carlos Carpio and David Hughes
Community Development, volume 38, issue 3 (Autumn 2007) pp. 77-89

Chapter 7

Options for the Economic Health of Farmers, Farmers Markets and Communities: Homebased Fruit and Vegetable Microprocessing
Sandra Bastin
Community Development, volume 38, issue 3 (Autumn 2007) pp. 91-98

Introduction
Sustainable Development: Connecting Agriculture and Community

Lionel J. Beaulieu and Jeffrey L. Jordan, Editors

Introduction

Environment, economy, and equity – the three words that have long served as the basic tenets of sustainable development in the U.S. Without question, attention to three core elements is due, in no small measure, to the seminal report of the World Commission Environment and Development (commonly referred to as the Brundtland Commission report) titled, *Our Common Future*. Released in 1987, the document defines sustainable development as "development that meets the needs of the present without compromising the ability of future generations to meet their own needs." The report outlined the three vital components of sustainability: (1) *environment*: the conservation and enhancement of the environmental and natural resources base; (2) *economy*: the development economies that focus on long-term economic stability and the wise use of resources; and (3) *social equity*: addressing the basic quality of life needs of all people (i.e., health, education, participation, equity) while remaining accountable to the critical needs of future generations (World Commission on Environment and Development, 1987).

In the context of agriculture, the quest to introduce and implement sustainable development strategies was intended to counter the expanded presence of large, industrial agricultural enterprises – firms whose activities were deemed to be harmful to human health and the environment. Horrigan et al (2002:446-48), for example, outlined a host of problems resulting from the practices of large-scale agricultural firms, including: (1) damage to the environment as a result of widespread use of fertilizers and pesticides; (2) soil/land degradation; (3) depletion of water resources due to reliance on irrigation; and (4) use of nonrenewable fossil fuels to support the food production system. The sustainable development effort began to take hold with the introduction of the Low-Input Sustainable Agriculture program in the 1985 Farm Bill, a program that became the Sustainable Agriculture Research and Education (SARE) program in the next (1990) Farm Bill (Berardi et al. 2011; Jordan and Constance 2008). As captured in the language of the 1990 bill, sustainable agriculture was conceived as an integrated plant and animal production system that would satisfy food and fiber needs while improving environmental quality, ensuring the efficient use of natural resources, and maintaining the economic viability of farmers and communities (Berardi et al. 2011; Chibulka 2009). In essence, SARE was designed to align with the three legs of sustainability referenced in the 1987 Brundtland Commission report – economic profitability, environmental stewardship, and improved quality of life (Tanaka and Bhavsar 2008).

Are there visible practices that have emerged as a result of the expanded interest and commitment to sustainable development? The simple answer is yes, and they include such on-farm strategies as crop rotation, no-till/low-till farming, soil management, nutrient management, rotational grazing and integrated pest management (Horrigan et al. 2002). For those with a concern with activities beyond the farm gate, some of the important efforts have included those designed to link producers and consumers through farmers' markets, reduce the presence of food deserts through the launching of urban gardens or community supported agriculture programs, engage a broad array of people and groups on local food policy councils, create local and regional food systems and promote honest dialogue on food system challenges and conflicts with the use of deliberative or study circle strategies (Gillespie 2010:10). These latter strategies embody what Lyson (2004) has coined as "civic agriculture" – agriculture that is embedded in the community and as such, helps build and maintain trust between producers and consumers (Carolan 2006).

Despite efforts to invest in the tripartite components of sustainability, Tanaka and Bhavsar (2008) found that the "quality of life dimension" of sustainability was given limited attention by researchers and practitioners engaged in SARE funded projects. For instance, they discovered that only 11 of the 174 projects (or six percent) funded by the Southern SARE over a span of 15 years were directly focused on community quality of life. It is this unbalanced project funding portfolio that prompted the Southern SARE program to forge a new collaboration in 2002, one that would link its sustainable agriculture development work with the community and economic development work of the Southern Rural Development Center. Since both regional entities operated in the same geographic area of the South, it appeared sensible to work in tandem to strengthen the community quality life portfolio of the SARE program.

The Role of the Community in Promoting Sustainability

With a commitment to partnering in place, the Southern SARE and SRDC launched the "Sustainable Community Innovations Grant Program" (SCIG) in 2002. Over the last decade, the competitive grants program has funded nearly 90 projects in communities across the South. The purpose of the program has been twofold: (1) invest in proposals that explore the variety of ways in which sustainable agriculture activities impact the well-being of communities – be it residents, economies, or institutions; and (2) fund projects that explore the manner in which community-level forces influence (in positive and negative ways) farm sector sustainability.

In order to broaden the focus of the SCIG proposals beyond those associated with SARE's traditional sustainable agriculture grant programs, applicants were urged to give attention to one or more of the seven "capitals" (natural, cultural, human, social, political, financial and built) associated with the community capitals framework (Flora and Flora 2013). In so doing, the SCIG program was able to invest in projects that incorporated a broader mix of factors that could impinge on the environmental, economic, and social well-being of agriculture and the community in which agriculture is embedded. Some of the important themes advanced in the SCIG program (and the types of capital to which they most closely relate) include the following:

Strategic Planning Efforts: Projects that engage local farmers, businesses, government, civic leaders, and others in the development and implementation of a strategic plan that gives careful consideration to sustainable community and agriculture strategies (natural, human, social, political).

Integration of Agriculture into the Local Economic Development Plan: Initiatives that link local agriculture/value-added activities with the comprehensive economic development strategies and activities of the local community or county (natural, cultural, social, political, financial).

Food Assistance/Security Activities: Programs that enhance food access and food security to low-wealth people and community via the development of innovative local food systems (human, social, financial).

Niche Food and Agriculture-Related Businesses: Activities by the community to strengthen, expand, and/or create new agriculture/value-added businesses that reduce economic leakages or expand the availability of locally/regionally produced foods. These include entrepreneurial initiatives designed to facilitate the development of entrepreneurs engaged in value-added agricultural activities (natural, cultural, human, social, political, financial and built).

Regional Agriculture-Related Economic Clusters: Initiatives that connect community/county governments, the agricultural sector, civic leaders, and others for the purpose of pursuing strategies that strengthen agriculture, agribusiness, and food processing activities on a regional (multi-county or multi-state) basis (natural, social, political, financial).

Civic Dialogue and Deliberation: Programs that engage residents in public deliberation or study circles on issues having important bearing on sustainable agriculture and/or sustainable community activities (natural, cultural, human, social).

Local Leadership Development: Projects that strengthen the ability of a broad-based group of local citizens to understand and promote sustainable agricultural/community approaches, policies and/or practices (human, social, political).

Overview of Book Chapters

The six chapters included in this book are not all studies that have been funded as a result of the Sustainable Community Innovations Grant initiative. However, they do capture the spirit and philosophy of the SCIG program. In its simplest term, the chapters are classified into two groupings. Three chapters focus on lessons learned from the community development literature that can help shape sustainable agriculture strategies. A second set of chapters provides insights from the sustainable agriculture literature that can contribute to the manner in which community development activities are pursued.

In the first category, Brasier, Goetz, Smith, Ames, Green, Kelsey, Rangarajan and Whitmer (Chapter 5: *Small Farm Clusters and Pathways to Rural Community Sustainability)* draw lessons from the industry cluster research and apply them to small farm clusters in the Northeast region of the country. Cluster are seen as groups of businesses that are part of the same industry, share common characteristics, coordinate some parts of their business activities but compete in others, and are geographically located near one another. Drawing on data gathered from 51 key informant interviews drawn from seven small farm clusters in the Northeast, the authors provide important perspectives with regard to structuring regional economic development efforts that lend valuable support to local agricultural clusters. The key to success, the authors argue, is the building of strong active networks among these farming clusters. These agricultural clusters are instrumental in promoting the economic viability of many farms, expanding access to affordable working farmland, building food processing facilities for use by small scale farmers, and finding specialty markets for their products. They also note the vital contributions that farm clusters make in building relationships with a variety of community groups and institutions, such as expansion of local tourism activities, the strengthening of voices/visibility of under-represented groups in the community, and imparting new knowledge to local youth and adults on agriculture's contribution to the region's economy, history, and culture.

Chapter 3 by Morton and Miller (*Connecting Sustainable Development to Rural Development: The Case of Pasture-Based Dairy Grazing*) devote attention to the technical and social infrastructure systems needed to create new rural economic opportunities. Focusing on a group of Ohio dairy graziers, they explore a grassroots ecosystem management model (GREM) that actively works to build broad-based civic engagement on the part of the local community. As the authors note, one of the important goals associated with GREM is to "engage citizens in public and private decisions that support humans *with* nature relationships rather than humans *over* nature." Part of the process involves the evaluation of scientific reports and the tapping of technical assistance. But these scientific assessments are then coupled with local knowledge and experiences – a process that allows the community to learn together in formulating sound and innovative decisions regarding the enhancement of the economic vitality of the farming community. The key community development principle underlying the GREM effort is the desire to strengthen "community agency," the set of horizontal relationships that help build trust and cooperation among a diverse group of local residents and that engages them in collective decision-making.

Wright, Score and Conner (*Food System Makers: Motivational Frames for Catalyzing Agri-Food Development through Multi-Stakeholder Collaboration*) in Chapter 3 note that multi-stakeholder collaborations is becoming an increasing important component of work in the community and economic development field. It calls for an integrative approach – one in which a diversity of partners are invited to the table and are given an authentic voice in shaping group decisions. The authors examine the Food System Economic Partnership (FSEP) in southeast Michigan, a five-county economic development partnership whose goal is to catalyze change in the region's agri-food system. They explore whether "agrarian fundamentalism" or "civic agriculture" serves as the central factor motivating stakeholders to take part in the FSEP. They find that 60 percent of the FSEP members embrace the "agrarian fundamentalism" frame, one which gives primacy to agricultural productivity, including support for value-added agricultural activities. The remaining 40 percent of the members subscribe to the "civic agriculture" frame, one whose concerns center on relationship building, social equity, and smaller scale food production enterprises. In the community development vernacular, the former focus reflects the "development in the community" mindset while the latter incorporates the "development of the community" approach (Robinson and Green 2011).

The second set of chapters– lessons from sustainable agriculture – focus on local food systems and how these assets contribute more broadly to community development. Hultine, Cooperband and Gasteyer (Chapter 6: *Linking Small Farms to Rural Communities with Local Food: a Case Study of the Local Food Project I Fairbury,*

Illinois) present a systematic evaluation of a local food system project in Central Illinois. The case study examines the creation of an indoor farmers' market by a group of local farmers inside an independently owned supermarket in the community. The authors suggest that the success of this innovative approach to developing a local food system is due to the existence and development of social capital – strong ties and relationships between producers and community members. In the end, they find that this type of local food system project offers rural communities valuable opportunities to improve local markets for farmers, build new relationships with their local businesses, and establish or expand connections with the community at large.

Consistent with the Hultine et al. discussion of local foods systems, Chapter 1 by Robinson, Robinson, Carpio and Hughes (*Linking Sustainable Agriculture and Community Development: the Low Country Food Bank's Use of Locally Grown Foods*) examines the use of locally grown foods by a food bank in South Carolina. The authors note that food banks are rarely perceived as contributors to local community economic development activities. More often than not, they are seen as conduits for providing emergency food assistance to individuals. But the authors carefully analyze the efforts of the Low Country Food Bank (LCFB) which launched the "Growing Foods Locally" (GFL) program in four counties in South Carolina. What the researchers discover is that GFL contributes to the local economy by helping increase profits to participating farmers. Moreover, it creates new food distribution and marketing channels and enhances the quality and nutritional value of food that the bank is able to distribute to a number of nonprofit agencies whose clients are of low-wealth.

The fourth chapter by Bastin, a study that focuses on efforts to broaden the products and market opportunities for Kentucky farmers (many who were formally engaged in the tobacco production), describes how state agency, higher education, and nonprofit representatives worked in partnership with state legislative leaders and farmers to support passage of a farmer friendly bill in their state legislature. House Bill 391, passed in 2003, allows Kentucky farmers to produce and sell valued-added home canned foods from their on-farm operation, farmers' markets, or certified roadside stands. Bastin's article traces the impact of this legislation on the farm sector. She finds that nearly 300 farms are now selling value-added products that contain fruits, vegetables, or nuts grown on these farms, resulting in increased profitability for these enterprises. As Bastin notes, the success of the Kentucky program is spurring other states to take a hard look at rewriting legislation that will allow home-based processing of jams and jellies by farmers.

Concluding Comment

When viewed as whole, these chapters incorporate several elements that are core to sustainable development. Some focus on ways to carry out agricultural activities that are both scientifically sound and dedicated to the protection of local natural resources (such as the article by Morton and Miller on "grass roots ecosystem management"). In other cases, the intent is to give voice to a wide array of local people, governments, organizations, and agencies who are dedicated to addressing the needs of agriculture and the community in a fair and balanced manner. Wright et al.'s focus on multi-stakeholder collaboration, and Brasier et al.'s efforts to reach out to groups often excluded from local decision making (such as Hmong, Hispanic, and women small farmer groups), are two articles that clearly come to mind. Still others are trying to find innovative ways to strengthen the economic health of the farm sector while simultaneously benefiting the community at large. Hutline's exploration of the benefits accruing to area small farmers as a result of the creation of a farmers market in a local supermarket (which offers local consumers access to healthy, fresh farm products), Bastin's study of the economic benefits to farmers as an outgrowth of new farmer friendly legislation in Kentucky (which allows a diversity of value-added products to be marketed through local channels), and Robinson et al.'s assessment of an area food bank's efforts to purchase locally grown foods so as to improve the nutritional health of low-wealth families and individuals in the locality, are three prime examples.

But, when all is said and done, the threads that weave our chapters together is the commitment to a building and expanding the community capital resources that have important bearing on the sustainability of agriculture and the broader community of which it is a part. Certainly, the success of the agriculture/community partnerships is rooted in one critical ingredient – "social capital" (Putnam, 2000). To be effective over the long-term, sustainable development depends on a network of people, drawn from a wide array of interests, who have a strong trusting relationship with one another, and who are willing to work together in responding to the economic, environmental, and social challenges facing agriculture and community alike. At the same time, it is important to appreciate and pursue strategies that work to strengthen the stock of the other six community capitals. It is balanced investments in all seven community capitals that will facilitate the emergence of what Luloff and Swanson (1995:352) have been termed "community agency" – the ability of local people to act in a proactive manner in managing, utilizing, and enhancing local resources. With the emergence of "community agency," an important step in the pursuit of a sustainable future for both agriculture and community is possible.

References

Berardi, G., R. Green and B. Hammond. (2011). Stability, sustainability, and catastrophe: Applying resilience thinking to U.S. agriculture. *Human Ecology Review* 18(2): 115-125.

Carolan, M. S. (2006). Social change and the adoption and adaptation of knowledge claims; whose truth do you trust in regard to sustainable agriculture? *Agriculture and Human Values* 23: 325-339.

Chibulka, D. A. (2009). Sustainable agricultural practices in the United States and other post-industrial countries. Pp. 81-94 in L. E. Phoenix (editor), *Critical Food Issues: Problems and State-of-the Art Solutions Worldwide (Volume 1)*. Santa Barbara, CA: ABC-CLIO

Flora, C. B. J. L. Flora. (2013). *Rural Communities: Legacy and Change* (4th edition). Boulder, CO: Westview Press.

Gillespie, G. W. (2010). 2009 AFHVS presidential address: The steering question: Challenges to achieving food system sustainability. *Agriculture and Human Values* 27: 3-12.

Horrigan, L., R. S. Lawrence and P. Walker. (2002). *Environmental Health Perspectives* 110 (May): 445-456.

Jordan, J. L. and D. H. Constance (2008). Sustainable agriculture and the social sciences: Getting beyond best management practices and into food systems. *Southern Rural Sociology* 23(1): 1-22.

Luloff, A. E. & L. E. Swanson. (1995). Community agency and disaffection: Enhancing collective resources. In L. J. Beaulieu & D. Mulkey, *Investing in People: The Human Capital Needs of Rural America* (Chapter 16, pp. 351-372): Boulder, CO: Westview Press.

Lyson, T. A. (2004). *Civic Agriculture: Reconnecting Farm, Food and Community.* Medford, MA: Tufts University Press.

Putnam, R. D. (2000). *Bowling Alone: The Collapse and Revival of American Community.* New York: Simon and Schuster.

Tanaka, K. and V. M. Bhavsar (2008). The role of Southern SARE projects in enhancing the quality of life in rural communities in the South. *Southern Rural Sociology* 23(1): 23-46.

World Commission on the Environment and Development (Brundtland Commission). (1987). *Our Common Future.* Development and International Co-operation: Environment. Oxford: Great Britain: Oxford University Press (December).

Small Farm Clusters and Pathways to Rural Community Sustainability

Kathryn J. Brasier, Stephan Goetz, Lindsay A. Smith, Molly Ames, Joanna Green, Tim Kelsey, Anu Rangarajan, and Walt Whitmer

Industry "clusters" are groups of businesses that are part of the same industry, share a common characteristic, coordinate some parts of their business activities but are competitive in others, and are geographically located near each other. Recent research has recognized the importance of industry clusters to regional development, competitiveness, and innovation. In particular, clusters are promoted as innovative models for rural communities in the face of globalization. Our research takes several lessons from industrial cluster research and applies them to clusters of small farms in the Northeast United States. As part of an on-going research and extension project, we are investigating the processes by which agricultural clusters form and the methods by which cluster members communicate, share and create knowledge, innovate in management, and relate to their surrounding communities. In this article we examine the ways in which agricultural clusters lead to more viable farms and sustainable rural communities. Data for this article are drawn from 51 key informant interviews from seven small farm clusters in the Northeast. Our findings are useful for cluster members and communities interested in promoting clusters.

Industry clusters have been defined in regional economics, urban growth and geography literatures as geographic concentrations of firms or businesses in a particular field that cooperate, share local resources, exchange knowledge, and also compete (Bernat,1999; Goetz et al., 2004; Porter, 2000). The importance of clusters and the agglomeration benefits they provide are recognized both by economic developers and policy makers (NGA, 2002; Porter, 2004). Building agglomeration economies is increasingly a strategic choice among economic development practitioners.

The benefits of clustering accrue primarily to the cluster as a whole and secondarily to the individual firms making up the cluster, and their customers. This is important because clusters can create a public value that in turn potentially justifies public interventions, including subsidies. Clusters are increasingly seen as key to the creation and exploitation of regional innovation and competitiveness. Because clusters are tied to specific places (regions), the benefits they create spill over into the wider community in which they are located.

Community benefits identified in previous studies include increased economic growth, jobs, wages, and worker training opportunities. These benefits have been documented for industries such as software development in Silicon Valley, chemicals in northern Ohio and the automotive industry in the I-75 corridor. Few if any studies examine the potential benefits of agricultural clusters in general and for rural communities in particular. Despite the enormous popularity of the cluster concept, "...in some respects the rush to employ 'cluster ideas' has run ahead of many fundamental conceptual, theoretical and empirical questions" (Asheim et al., 2006, p. xvii). Systematic studies of how clusters form, how they can be nurtured, and how they lead to community-level benefits are virtually non-existent. In this paper, we attempt to elucidate some of the conceptual and theoretical issues surrounding clusters by specifically examining case study data that links agricultural clusters to the growth and sustainability of their surrounding communities.

Industry Clusters

Clusters are defined as "geographic concentrations of interconnected companies, specialized suppliers, service providers, firms in related industries, and associated institutions (e.g., universities, standards agencies, trade associations) in a particular field that compete but also cooperate" (Porter, 2000, p.15). Clusters include but are not limited to supplying firms, firms that make the product, firms that market or sell the products, those that increase and share knowledge pertaining to the product, and firms that help to create the rules and regulations concerning the product and its production. Each of these cluster members has particular objectives and goals specific to their own business; however, they share an interest in the success and further development of the industry as a whole.

The networks that coordinate transactions among firms within the cluster vary from formal to informal. These networks not only coordinate activity, but manage competition across the firms to "bind buyers and sellers in customary ties of familiarity and reciprocity. This, in turn, encourages forms of cooperation that often enhance the comparative economic advantages of particular industrial localities" (Scott, 1992, p.223).

Clusters provide a number of collective benefits to their members. First, geographic proximity and coordination allow member firms to form a critical mass to purchase services that no individual firm could afford to maintain in-house. For example, they may jointly contract for marketing, research and development services, worker training, etc. (Scott, 1992). Further, coordination of activities allows clustered firms to respond more flexibly to changing market conditions. This can be particularly crucial in new and emerging industries and market areas. This constant interaction creates innovations and forces individual firms to differentiate themselves to remain competitive with their rivals (Porter, 2000).

As a result of this geographic concentration of firms, infrastructure and ancillary support services form to serve these businesses (Dawkins, 2003). In addition, workers with skills appropriate to the industry locate in the area, to improve their employment prospects (Cortright, 2006). Worker training programs are often developed cooperatively with cluster firms, employment centers, colleges and universities, and local labor unions. These educational efforts and infrastructure development is frequently part of a concerted effort by local governmental and economic development agencies (Dawkins, 2003).

In clusters, firms *compete cooperatively* and they *cooperate competitively*. This delicate balance is achieved only by organizational and communication patterns among cluster members that have the potential to influence firms' competitiveness, management techniques, innovation capacity, flexibility, environmental impacts, social support, and community relations. These benefits extend beyond the individual producer to the cluster as a whole, creating an environment in which collective learning and sharing of resources is fundamental to the business philosophy.

This network or organizational ability is crucial to the success of clusters (Scott, 1992). "[G]eographic clusters facilitate intralocal information flows and help to spread ideas and innovations especially when firms share similar local business cultures, interact in local business organizations, and share similar local or national norms and values...." (Dawkins, 2003, p.144).

The crucial role played by networks has been identified within agricultural work, particularly as related to adoption of technologies and growth in alternative agriculture. Trusted individuals, who are similar to each other, are likely to have a greater effect on the integration of information and adoption of practices (Rogers, 2003). This trust is most often built through repeated interaction within local networks (Coughenour, 2003; Warriner and Moul, 1992). Clusters encourage the growth of these local networks that can induce change in management behaviors. Research within the sustainable agriculture movement has emphasized the role that private networks play in disseminating local knowledge among producers to help them learn about techniques, technologies, practices and environmental idiosyncracies that will make them more successful (Hassanein and Kloppenburg, 1995). This is particularly true for those learning a new or non-traditional approach to farming, for which few if any formal resources may be available. Further, the trend toward increasingly privatized and "commodified" information in agriculture suggests that these kinds of networks will become especially important for small- and medium-scale farmers who lack the resources to access needed management information (Goe and Kenney, 1986; Wolf and Wood, 1997).

In summary, clusters share a number of characteristics. They are concentrations of firms or businesses that (1) are located in relatively close proximity to each other; (2) compete with each other in similar markets; (3) cooperate with one another to enhance both technical skills and market access; (4) support, through social networks, the growth and development of individual businesses; (5) share common inputs such as labor with specific skills; (6) recruit support industries based upon the concentration of firms in an area; (7) benefit mutually from new knowledge generation that is location specific; and (8) work together to respond to new market needs or societal demands.

Impact of Clusters on Surrounding Communities

With few exceptions, the discussion of clusters' impacts on surrounding communities has been largely theoretical. Here we review potential impacts discussed in literature that apply to agricultural clusters. Impacts can be economic, social, environmental or cultural.

Economic Impacts

Specific economic benefits include lower transportation costs for firms in the community because of input and output market agglomeration, higher worker productivity and skills, as well as spillovers of knowledge and accelerated spreading of innovation (Bernat, 1999). Employment growth and retention are increased and risks are pooled more effectively.

Gabe (2004) finds that clustered firms invest more funds in worker training than non-clustered firms, even in rural areas. Gibbs and Bernat (1998) show that rural firms in clustered industries pay higher average wages than do firms that are not clustered. More generally, clustering affects and has implications for state-level workforce training efforts, and the types of curricula and programs offered by local universities and community colleges (Aspen Institute, 2002; Herr, 2003; Marano, 2004).

Existing clusters can help to support entrepreneurial activity within communities.

Experience gained and networks created by working with clustered firms can be used to start new related businesses (Cortright, 2006), which in turn support local economic growth through import substitution and other innovative activities. More generally, clusters that produce export items are a critical part of a community's economic base. Through direct and indirect multiplier effects they lead to the generation of new income and employment locally. Moreover, clusters can interact across sectors and industries to provide further leveraged benefits. For example, the Finger Lakes Wine Cluster in New Yrok is a vital component of tourism in that region – and increasingly so as heritage tourism becomes more important. Likewise, when the sugar cane industry relocated from Salt Lake City, Utah to Idaho seemingly tangential sectors such as logistics, marketing and finance were impacted adversely at least in the short-term.

Yet economic benefits are not universal or automatic. Agglomeration diseconomies may arise (Gibbs and Bernat, 1998) and industry clusters do not necessarily always boost rural job and wage growth. Instead, this depends on the industry and the community. Barkley and Henry (2002) find that the costs of pursuing cluster development may exceed the benefits to communities and states, depending on the industry.

Social Impacts

In terms of social spill-over benefits, significant research has been conducted on the impact of farm characteristics on community well-being, following Walter Goldschmidt's seminal work, *As You Sow* (1978). Although the empirical findings are mixed, this line of research indicates that communities with large-scale farms that rely heavily on hired labor tend to have lower well-being (poverty, inequality, income) indicators. Later work has indicated that small-scale farms have a similar negative impact (Lobao, 1990). This research is similar to that on non-agricultural communities, where the presence of small- to medium-scale businesses creates higher levels of social capital and commitment to community and better overall conditions (Tolbert et al., 1998). Nevertheless, this work does not address the potential unique contributions of *clusters* of agriculturally-related businesses on community sustainability.

Because the crucial element of clusters is the underlying network through which information is shared, we draw on literature that examines community well-being as related to the qualities of community networks. Theoretical and empirical work emphasizes the importance of a committed network of community members who work toward improving the community's quality of life, economy and natural environment (O'Brien et al., 1998). This work has identified a number of characteristics of these networks that lead to better outcomes, including: the diversity of members within the network, the strength of the horizontal interaction among members of the network, and the number of vertical ties to resources outside the network (Flora and Flora, 1993). Lee and colleagues (2005) find that social networks within communities can increase the visibility of a particular region for the purpose of "selling local products, attracting outside resources, and engendering a stronger sense of identity" (p. 274). Consequently, the qualities and characteristics of the relationships within the clusters' underlying social networks become crucial for understanding the potential community effects.

Environmental Impacts

Clusters of farmers may improve biological and environmental management practices on farms, reducing the local impact on natural resources. Several examples exist of farmers and other community members within particular geographic locations working together to modify management practices for environmental improvements, including decreased use of Atrazine (Salamon et al., 1998) and the 'Healthy Grown' potato.[1] The interaction of these farmers and related organizations creates openings for building trust, sharing innovations, and discussing how best to blend agricultural and natural resource management goals.

Cultural Impacts

Evidence of cultural or heritage-based cluster benefits is largely anecdotal. Hilchey (2006, p. 204) notes that:

> …an alternative path to the future of [these] traditional commodities may be to capture and market their rich regional heritage as well as the enormous contributions they make to American foodways and working agricultural landscapes. National, state and local "heritage areas" already exist, building educational tourism around sites of significant events and developments in American history, such as wars, industries, canals, women's rights, slavery, and the like.

In summary, to the extent that the path to regional economic competitiveness leads through innovative, competitive and well-functioning industry clusters, the link from clusters to sustainable communities is clear. The largest benefits likely emerge when all four dimensions – economic, social, environmental and cultural – are allowed to mutually reinforce one another. To our knowledge, there are no studies that examine the community impact of a cluster of farms and related agricultural businesses and support agencies. This article explores the potential pathways through which clusters impact local community well-being. We draw from a study of seven agricultural clusters in the Northeast U.S. to delineate these pathways.

DATA AND METHODS

Identification of Agricultural Clusters in the Northeast U.S.

Agricultural clusters in the Northeast region of the United States were identified through research and discussions with key individuals working in the agricultural sector at both a statewide and local level. Eight clusters were selected, but one chose not to participate in the project. Clusters were chosen to represent a range of reasons for organizing, including around markets and market opportunities, identity, and commodity; location; and stage of development (Table 1).

Table 1. Clusters Selected and Their Characteristics

Cluster	State	Type	Number of Interviews
New York Certified Organic	NY	Organic	6
Mass. Hmong Farmers	MA	Ethnic	6
North County Dairy Viability Initiative	NY	Dairy	5
Pennsylvania Women's Agricultural Network	PA	Gender	6
Chesaspeake Fields	MD	Specialty grains	17
New Farmer Development Project	NY	Minority	7
Tuscarora Organic Growers	PA	Organic	4

Methods

Multiple key informant interviews were conducted to develop a comprehensive account of each cluster. Within each cluster, individuals were identified through document research and snowball sampling. Key informants were chosen based on their knowledge of the cluster and their position within the cluster. Key informants included formal leaders and board members, informal leaders and highly engaged participants, external agency and organizational representatives working with the cluster, and members of the surrounding community aware of the clusters but not formally affiliated.

In all, 51 interviews were conducted with 61 key informants during the spring and summer of 2006. The number of interviews conducted varied by cluster (see Table 1). All interviews were conducted using the same structured interview schedule. The questions covered issues related to the individual's connection to the cluster; the history and formation processes of the cluster; identification as a 'cluster'; the organization and communication processes used within the cluster; the types of exchanges (social, informational, financial) that occur between cluster members; the perceived benefits and challenges of cluster participation to individual cluster members, the cluster as a whole, and to the surrounding community; and participants' perceptions of the future of the cluster. This schedule was developed jointly by the multidisciplinary research team.[2]

Three members of the project team conducted the interviews. To ensure consistency and continuity, the same interviewer conducted all the interviews within a cluster. Further, all project team members jointly developed the interview protocols and the instrument, and therefore understood the intent of the processes and questions. All team members participated in multiple sessions both before and during the interviews to practice interviewing techniques and review interview protocols.

Of the 51 interviews, 31 were conducted in person and 20 via telephone. All but four interviews were conducted one-on-one. The group interviews occurred for multiple reasons, including the need for a translator and cultural sensitivity. Two group interviews occurred because the researcher arrived and found that multiple cluster members were waiting, eager to talk with her.

Interviews were audio-recorded and the tapes transcribed verbatim. The transcriptions were checked for accuracy by both the interviewer and the key informant. All members of the multidisciplinary research team were engaged in the process of identifying emergent themes. Each interview was read by at least two team members; no researcher read an interview which he or she conducted. Each researcher developed a list of themes based on the 20 interviews read. These lists were collated and, after significant discussion, merged into a coding scheme. The resulting coding scheme highlighted the main 'big themes' that both emerged from the data and drew on the literature related to clusters. Sections within the transcripts were coded according to this scheme. Data were managed through the use of qualitative analysis software (NVivo N6©). Transcript sections were then sorted by theme. For the purposes of this manuscript, summary documents and interpretive memos were created that organized the text specifically for the codes related to community impact.

RESULTS

Cluster Summaries

New York Certified Organic (NYCO): NYCO started in 1994 as the New York chapter of the Organic Crop Improvement Association (OCIA), an organic certifying agency. In 2001 the group split from OCIA and officially became known as "NYCO." They stopped certifying, but remained focused on developing an educational and community network. NYCO participants are located in the central New York region. This group has roughly 100 participants, 20-25 of whom regularly attend events. Within NYCO there are both farmers who buy organic grain (dairy farmers) and farmers who sell organic grain, as well as a grain miller. Members are encouraged to *think like a cluster,* not like autonomous competitors. NYCO members articulate that the cluster will benefit as a whole if prices are kept reasonable for both producers and buyers of grain.

A key function of this cluster is to mentor and provide support to farmers considering the transition to organic. This support takes the form of formal education, informal sharing of ideas and experiences, and shared investment in on-farm research efforts. A high priority is also given to building supportive relationships with fellow producers, business, and resource people. It is a purposeful alternative to what they view as the culture of isolation and competition of the mainstream agriculture community. NYCO leaders have very consciously nurtured a culture of community by modeling values such as cooperation, equity, transparency, honesty, trust, mutual respect, valuing and enjoying each other as whole people (not just economic partners), and shared decision-making. This environment creates many opportunities for creativity and innovation among participants.

New Farmer Development Project (NFD): The New Farmer Development Project grew out of a series of meetings discussing ways to increase participation in the New York City Greenmarkets in 1999. These meetings were highlighted in a Spanish language newspaper, resulting in tremendous interest within this immigrant population.

Members share a common heritage of agrarian skills and a desire to take advantage of their proximity to urban and growing ethnic markets. The group consists of Hispanic, immigrant farmers living near New York City; a core group of 20 farmers serve on the steering committee. Three staff members organize activities in response to member requests. Because many of the farmers originally came from semi-tropical climates, they had difficulty adapting their skills and knowledge to the northeastern U.S. growing environment. Members openly share their experiences, and, as a result, have been able to find unique adaptations to technical production problems.

Benefits to individual members include access to information, advocacy, and social and emotional support. Staff members act as buffers between the immigrant farmers and outside agencies so that questions can be asked without fear. Members report gaining confidence in their skills and in their future because the NFD project has developed a critical mass of successful growers and marketers.

Chesapeake Fields (CF): Chesapeake Fields is a group of growers, processors, and community members focused on specialty grain (wheat, corn, soybeans) production in Maryland. This cluster has three formal entities: Chesapeake Fields Cooperative, Chesapeake Fields LLC, and Chesapeake Fields Foundation. Farmer members of the Cooperative supply raw product to the LLC and in turn receive a portion of the proceeds. Investors (farmers and non-farmers) in the LLC own the infrastructure for the processing facility. The nonprofit Foundation is a broad group of community stakeholders supportive of the endeavor. Chesapeake Fields developed in response to community residents' concerns over the rapid loss of productive farmland during the 1990s. Their goal became 'preservation through profitability,' focusing on value-added market opportunities.

Cluster members share production techniques to improve the quality of their crops. They also share information regarding marketing opportunities. The cluster organizes research and business planning activities that have resulted in innovation for individual businesses, as well as new products, markets and business structures. The entire value chain is represented across the three entities: non-farm rural residents and business owners; farmers; government agencies, politicians, and other supporting organizations. Although the farmers grow and sell the same products, and may in some cases be in direct competition with each other, they sell only a portion of their products through the LLC and co-op. Members share transportation and storage facilities, as well as collaborative marketing of the group and their products.

Pennsylvania Women's Agricultural Network (PA-WAgN): PA-WAgN, formed in 2003, consists of approximately 850 women farmers, educators, agribusiness owners, and agricultural professionals across Pennsylvania. PA-WAgN is modeled after networks in Maine and Vermont. The structure consists of a steering committee of women involved in agriculture across Pennsylvania, plus staff and faculty located at Penn State University. Each region of the state has developed small cells of women farmers who communicate and organize activities informally.

PA-WAgN provides educational, leadership and networking opportunities. Participants share information and support around issues impacting their roles as women, as farmers and as family and community members. The participants in PA-WAgN compete with each other in that most sell products directly to consumers, but in geographically dispersed markets. The group agreed that demand for their products so far exceeds the supply, creating opportunities for collaboration. Members articulated that increasing awareness within the general consumer population of small, diversified farms benefited all growers. Additionally, through education to understand their costs and market potential, many members felt that prices for all growers were higher. The group's activities create a supportive atmosphere that is conducive to risk-taking and entrepreneurialism.

Hmong farmers: This is a group of eight Hmong growers who produce Asian vegetables on a field owned by a Portuguese-American dairy farmer in Massachusetts. In 1984, an elderly Hmong woman asked if she could grow a garden behind the farm's dairy barn. In the second and succeeding years, the opportunity grew by word of mouth. This group of eight farmers is a subset of 60 Hmong families who grow produce non-commercially on the farm's land.

The eight commercial growers produce and market their products independently through farmers markets in about a one-hour radius of the farm. The growers meet regularly as a group, with leadership from the farm owner, to: make decisions about land preparation, allocation of plots, and land improvements; coordinate marketing so that they are not competing against each other at a given market; access information, training and other resources; and access research and demonstration results. As of January 2007, the group became a formal committee (with the assistance of Heifer International) to organize group activities and allocate shared resources.

Hmong cultural traditions shape the ways in which the group operates, including: strong familial and community bonds; distrust and fear of outsiders, agencies and government; importance of showing respect according to Hmong cultural norms and traditions; and a hierarchical social order. In addition, language barriers often make communication with the farm owner and other outsiders problematic.

The eight commercial growers get the support they need to overcome language and other cultural barriers to market their products and make an income from farming. The farm owner connects the growers to external agencies and resources, and protects them from those who might exploit, misunderstand, or disrespect them. They are encouraged to form their own relationships with these organizations when they are ready. The sixty families growing their own fresh produce represents a major contribution to food security and a culturally appropriate diet for the local Hmong community. All the growers are empowered through the opportunity to farm and produce their own food, which is highly valued in Hmong culture. They are able to connect with their traditions and their memories of life before coming to the US, and there is some indication of improved mental well-being among the growers.

North Country Dairy Viability Initiative (NCDVI): NCDVI aspires to include participants throughout the entire "value chain" of the dairy industry in six northern counties of New York, including dairy farmers, manufacturers, lenders, and cooperatives, plus representatives of economic development entities and Cornell Cooperative

Extension. NCDVI formed as a response to concerns about the viability of the dairy industry in the face of increasing national and international competition. NCDVI was launched formally in 2001 with a series of four "listening meetings." These initial meetings engaged hundreds of participants, up to half of them farmers. Later activities engaged a dwindling number of farmers. Representatives of the dairy manufacturing sector have been more consistent participants.

NCDVI provides a forum for communication, education and relationship-building across sectors of the dairy industry with the aim of increasing the industry's viability. A steering committee (originally 20-25 volunteers) organizes activities at the regional level; county committees coordinate activities more locally, although only a few county committees have been active. The steering committee is now down to 6-8 individuals, who have not been meeting regularly. There is currently no budget to hire a coordinator, nor has anyone volunteered. The future of the NCDVI is very much in question at this time.

NCDVI made a conscious effort to help participants learn how to think like a cluster. NCDVI provided opportunities for sharing and appreciating the perspectives of other industry participants, shared decision-making across sectors, and open sharing of information and ideas. They particularly tried to cultivate participation from small, mid-size and large farms. As a cross-sector initiative NCDVI is uniquely positioned to promote innovations that require collaboration across the industry. However, efforts to develop such innovations have largely failed because of divisions between farmers and manufacturers in an extremely competitive economic environment. Participants are wary of each other, and cultivating a sense of shared interests, trust, and cooperation has been extremely difficult.

Tuscarora Organic Growers (TOG): TOG started in 1988 as a group of seven farmers in south central Pennsylvania who began co-marketing their products. By working together, the farmers were able to mitigate production costs and coordinate crop production for direct market needs, primarily to the Washington DC metro area. Today there are over 25 certified organic growers who sell over 50,000 cases of produce to restaurants, retailers and food stores in Pennsylvania, Maryland, and Washington D.C. year round.

TOG is first and foremost a marketing cooperative, with farmers as primary members. TOG farmers coordinate crop production, and work with the cooperative staff to manage sales, marketing and shipping of the produce from their distribution center. Although TOG is designed to minimize competition, farmers are not required to commit all of their produce to the cluster, and there have been a few instances in which farmers have sold their surplus in the same market as TOG. Conflicts are resolved and decisions made through open discussion and voting. While there is an elected Board of Directors who make most of the decisions, all meetings are open to all members. In the winter many of these meetings become open forums and "crop improvement meetings" where the growers can share information about farming. Members at times also share the costs of inputs such as potting soil or boxes. Overall, TOG links producers and consumers, manages competition, and has played a key role in the growth of direct market opportunities in the Washington D.C. area.

Pathways to Community Sustainability

The following community benefits of agricultural clusters were identified by the key informants, some of which match well with the industry cluster literature, while others are not mentioned previously and are specific to agriculture.

One of the main expected outcomes of clusters is economic growth for the surrounding area. In some clusters, this resulted from the formalization of business ties and industry growth. For example, the formation of Tuscarora Organic Growers resulted in the development of a distribution center that is "one of the largest employers" in the township. In other clusters, job growth was not mentioned directly but higher income for the individual growers and potential multiplier effects were described by several cluster members. This is particularly important in the clusters with new, immigrant, or transitioning farmers, where the farm became a primary source of family income.

Another feature is growth in the number of firms directly and indirectly related to the industry. This can be seen particularly with NYCO, where the cluster is at least partially responsible for rapid growth of the organic dairy and grain industry in central New York. Further, the cluster's activities enabled farmers to transition to organic more successfully than they would have on their own, increasing farm income. NYCO has also spawned at least two new agribusinesses, an equipment dealer specializing in cultivation equipment and expertise specific to organic production methods and a new local grain mill that specializes in organic grain seed and organic dairy feed mixes.

One anticipated outcome of clusters is development of infrastructure to support a growing industry. This infrastructure could not be supported or be viable unless multiple firms are present. In the case of agricultural clusters, one key infrastructure element is the availability and affordability of working farmland:

> If the farms disappear, we won't have this place any more…. So do others benefit from Chesapeake Fields trying to make sure that we keep our farms and to keep farmers profitable? Definitely. Definitely and you know how does it trickle down? Well farmers are consumers too… So they have their job and then it trickles all the way through the economy to all. Everybody would benefit from keeping our farms profitable and keeping our farmland operating. (CF)

Having a viable industry provides higher profits to the participating farmers, and brings optimism for the future viability of the industry which affects both farmers and local officials:

> I think when you look at it from the elected official's interest…, there is definitely a level of support and … an interest in seeing Chesapeake Fields be successful. There is an interest in retaining our farmland and making our farmland successful so that it can compete against the demands for developing farmland. (CF)

Cluster members also identified the development of facilities that specifically accommodate small farmers and those interested in small-scale food processing. One person associated with the Hmong group felt that: "There has also been certainly a rise in incubators and commercial kitchen kind of places … providing small business people and small growers with an opportunity for value added products." She attributed this growth to the rapid expansion of direct marketers like the Hmong, and the types of specialty products they provide. In addition, these farmers could earn even higher revenues if they could access such facilities and meet the demands of a year-round market.

Relatedly, working together across the value chain allows farmers to develop and maintain a market share that assures farm stability into the future:

> That is where we can be most profitable and make a name for ourselves. A niche, specialty, high end and high quality, good value product line. That is where we can make the most impact and dollars and have the great pay back through our farmers to make them as resourcefully stable as possible. Socially, financially, economically and generationally. (CF)

A potential 'spillover' effect of clusters is the improved viability of surrounding firms. One individual associated with the North Country Dairy Viability Initiative felt that even farms not directly participating in the Initiative increased their productivity because of the education and publicity generated by the Initiative's efforts:

> We look at the statistics when they come out and cow numbers are up and the milk was up and so I am thinking okay that this whole project must have benefited them because that was the whole idea of it. So they must have gotten something from this so that it helped them with their business.

Another potential impact on local communities is an overall increase in the market for a product, and resulting increased profitability of local firms. In the clusters focused on producing organic and/or local foods, this was clearly an identified benefit:

> I remember one of the farmers was almost angry at me.... This is not fair. This is not right. We had at least two or three farmers in the project [selling at the market]. You never saw so many faces at the market. He had a very good spot. They would be waiting for him and the strawberries and our farmers do not have it so.... He has to come to the point to understand that that was not a competition. It was creating more like demand for his stuff. (NFD)

Some of the less-discussed impacts of clusters are those directly related to community development. Building relationships across multiple institutions was identified as a key outcome of clusters that can lead to innovation and the growth of related industries:

> I have hired a number of students or recent graduates from [the local college] to do particular projects for us...You are building community relations there...one of the good things of him doing our research is [our] products are now being sold to [the] college because he simply said, I do research for this, and how come we are not doing this, and he went in to talk to the people who handle their student stores and their catering business and who ran their cafeteria and such. (CF)

This group also played a significant role in further developing the local tourism industry, by working directly with the tourism agencies and the farmers to identify unique farm-related visitor opportunities.

Because a number of the clusters worked specifically with under-represented or disenfranchised groups within agriculture, the clusters were particularly helpful in offering them opportunities to learn in a safe environment and increase their leadership and production skills. This empowerment of a community group gives them greater visibility, voice in community affairs, and ability to participate as community members: "My goal is to get them to know how to make decisions as a group. Actually make the decisions and not have one person making all of the decisions and doing everything" (Hmong). This empowerment can pave the way for other community groups: "I think they do benefit ... the community, not just the Hmong community. I think they do benefit to such like a Laotian or the Vietnamese community or the Chinese community." One of the ways in which these benefits occur is by legitimizing the culture in a new and unfamiliar environment: "It is more encouraging to see your products when you go to the farmer's market as compared to see that you have to be forced to try all of these new products....You come from a different background and you want to see things that that you are accustomed to eating" (NFD). This legitimacy was a particularly important result of the development of PA-WAgN:

> Now there is this statewide organization of women who are farming and I see their picture and other people see their pictures and hear about that it is going on in their life. Oh and it is not the Grange and it is not Young Farm Wives Group. It is like women who are saying that they are farmers.

As a result, women farmers have gained significant confidence in their skills and abilities to work within other community organizations, knowing that they are not 'alone.'

Another potential outcome of industry clusters is an increased awareness of the local industry and a statement about its importance for local identity. In areas that are historically, economically, and culturally tied to a particular type of farming, this keeps community members tied to the industry even in the face of change or conflict. In the North Country, one topic identified by the Initiative was that

> ...the non-farming community or the public needed to be educated...So we have opened up and have had farm tours and had like a family farm day... that kind of keeps the public informed of where the dairy industry is going and what the dairy industry is doing and how they do it and why they have to do things a certain way.

Relatedly, Chesapeake Fields offers educational opportunities to local residents, particularly youth, so they can learn about the local farming industry. They expect this to have significant impact on the future decision-making of the area's youth:

> ...that educational facility [is] where our youth can go in and find out how the food that they are eating in their lunches was grown here, produced here and how it did not just come from the shelves of the food store. It came from the farm and it has been processed, packaged and then purchased so that is kind of building the future so that our youth will also see the importance of that and be able to identify where everything comes from.

Another benefit identified by key informants is that of attempting to influence the policy environment in which the cluster functions. Chesapeake Fields, NCDVI, and NYCO specifically work with local, state, and national representatives to help them understand the industries, their local impact, and the challenges they face. In addition, cluster members gain experience, skills, and knowledge of the industry as a whole, and become experts who speak to groups at local, national, and international levels: "Thomas and John. Weren't they in Italy? ...And John speaks a lot. He gets around. I will see him speak to different...organic or some type of organizations in other states" (TOG). These skills can be particularly important in managing and negotiating relationships at a very local level:

> Some of these farms have fields that are surrounded by conventional farms growing genetically modified crops...How do you negotiate and work with your neighbors effectively so that you are planting in a different window or planting a different crop ...in order to avoid contamination? I do not think that there has been a case where a load has been rejected [because of contamination]...because they have been successful with working at it. (NYCO)

Clusters also give the non-farming community an "outlet" that encapsulates their interest in living in the region. This is particularly the case for Chesapeake Fields, where the development of the three organizations has provided ample opportunities to direct investment of money and energy in a shared goal and vision for the region. The advantage gained by the cluster is a coalescence of like-minded community members around a purpose. The group has provided the table around which people can gather to attempt to identify and solve problems.

> I think Chesapeake Fields is a tremendous advantage to the region and I think it has the potential to do some really good things to retain our culture and to retain our environment. To help us to retain all of the things...that brought many of us

> to this region in the first place. It is under very serious potential for changing and not for the better.

The structure of Chesapeake Fields encourages this shared goal:

> I find enjoyment bringing an idea to a hard live product. It is fun....It gives you a sense of professional and to a degree a personal accomplishment. Now the caveat is that... anyone involved in this organization cannot allow a personal agenda...overshadow or eclipse the other pieces and the other organizational agendas and goals.... Yes my mission is to go make money, but not at the sacrifice of the overall organizational goal or mission nor can it be for the institute, nor can it be for the cooperative.... It forces you to work more globally over your acting locally. That is not *a bad thing (CF).*

DISCUSSION

While clusters are seen as economic growth opportunities, very little research has documented the local impacts of such clusters. In this article, we describe the impacts identified by participants in agricultural clusters in the Northeast U.S. These results provide insight into the multi-dimensional pathways through which the networks that support the clusters can have far-reaching effects on the economic, environmental, social, and cultural characteristics of surrounding communities – characteristics which may lead to further development efforts.

Key informants in this analysis described economic impacts of clusters, particularly related to the sustenance and growth of viable farm and farm-related businesses and infrastructure (particularly working farmland). However, more discussion was given to the development of supportive networks that foster the growth of local entrepreneurs and other community endeavors. This was particularly visible in Chesapeake Fields and NYCO, where cluster leaders have purposely built strong, diverse networks that support such initiatives. These types of networks provide crucial information flows and a basis from which additional educational opportunities, innovation, and community efforts can be quickly developed (Korsching and Allen, 2005).

In the clusters focused around education and empowerment of non-traditional farming groups (Hmong, New Farmer Development, PA-WAgN), there is less clear connection to broader community development efforts. Instead, the pathway is through the development of skills within these groups, skills that can be used to improve profitability of individual farms and the voice of diverse groups within their surrounding communities. However, economic impacts have been felt within the local direct market farming economies of these communities as well.

North Country Dairy Viability Initiative is the one cluster we examine that is focused around a traditional commodity. The market conditions of the dairy industry, and the traditional antipathy between farmers and processors, have made efforts across the value chain very difficult to move forward. Although key informants did link improvement in individual farm viability to the efforts of the Initiative, few broader community impacts were identified. In comparison with the other clusters that have growing market opportunities, NCDVI was not able to establish a broad, diverse, and inclusive supporting network that fosters the development of trust, openness, sharing of information, and innovation.

CONCLUSION

This research provides several lessons for structuring regional economic development efforts that intend to support and nurture local agricultural clusters. First, the network of relationships that support the cluster is the central locus for deliberate effort. In particular, emphasis should be placed on forming relationships across the value chain, creating an environment in which open sharing of information is the norm, developing trust among all members, working toward a shared commitment to the 'cluster,' and recognizing both opportunities and limitations of individual businesses.

Second, the market and competitive context in which the farms and related businesses function will condition the types and qualities of interactions among cluster members, and will likely result in variable but deliberate efforts to create the network. It is also necessary to recognize market changes, and the impact these could have on the relationships within the network.

Third, it is essential not to overlook individual and group empowerment as fundamental elements of cluster growth. Growing confidence in individual abilities, knowledge of markets and governing institutions, and connections with other growers can lead to improved business viability, innovation, and ability to participate as equals in community affairs. Participation in these networks can offer practice in essential civic engagement skills (such as public speaking) and the development of knowledge that can be shared with others. Individuals with these skills are valuable community resources that can be drawn into related community efforts.

NOTES

1 The World Wildlife Fund, the Wisconsin Potato and Vegetable Growers, and the University of Wisconsin collaborated to develop bio-intensive integrated pest management practices and labeling for the 'Healthy Grown' potato (see http://128.104.239.6/bioipm/).

2 The disciplines represented within the project team include: agricultural economics, rural sociology, and horticulture. In addition, several team members are employees of (or have responsibilities to) their institutions' extension system

REFERENCES

Asheim, Bjorn, Philip Cooke, and Ron Martin. (2006). *Clusters and Regional Development: Critical Reflections and Explorations*. Oxon, UK: Routledge.

Aspen Institute. 2002. *Gaining Ground: The Labor Market Progress of Participants of Sectoral Employment Development Programs*, SEDLP Research Report No. 3, 63 *pp.*

Barkley, D. L., and M. S. Henry. (1997). Rural Industrial Development: To Cluster or Not to Cluster? *Review of Agricultural Economics* 19 (2): 308-325.

Bernat, G. Andrew Jr. (1999). Industry clusters and rural labor markets. *Southern Rural Sociology,* 15: 170-187.

Cortright, Joseph. (2006). *Making Sense of Clusters: Regional Competitiveness and Economic Development*. The Brookings Institution Metropolitan Policy Program.

Coughenour, C. Milton. (2003). Innovating Conservation Agriculture: The Case of No-Till Cropping. *Rural Sociology* 68:278-304.

Dawkins, Casey J. (2003). Regional Development theory: conceptual foundations, classic works, and recent developments. *Journal of Planning Literature* 18:131-172.

Flora, C. B. and J. Flora. (1993). Entrepreneurial Social Infrastructure: A Necessary Ingredient. *Annals of the American Academy of Political and Social Science* 529: 48-58.

Friedland, William H. (1984). Commodity systems analysis: an approach to the sociology of agriculture. *Research in Rural Sociology and Development* 1: 221-235.

Gabe, Todd M. (2004). Industry Agglomeration and Investment in Rural Businesses. *Review of Agricultural Economics* 27:89-103.

Gibbs, R.M., & Bernat Jr., G.A. (1997). Rural Industry Clusters Raise Local Earnings. *Rural Development Perspectives*, 12(3), 18-25. [Electronic Version]. Retrieved July 21, 2004 from http://www.ers.usda.gov/publications/rdp/rdp697/rdp697d.pdf

Goe, Richard W, and Martin Kenney. 1986. The Information Age: Implications for U.S. Agriculture. *Policy Studies Review* 6(2):260-272.

Goetz, S.J., M. Shields and C. Wang. (2004). Food and Agriculture Industry Clusters in the Northeast U.S.: Technical Report. Regional Rural Development Paper, The Northeast Regional Center for Rural Development.

Goldschmidt, W. R. (1978). *As You Sow: Three Studies in the Social Consequences of Agribusiness*. Montclair, NJ: Allanheld, Osmun and Co.

Hassanein, Neva, and Jack R. Kloppenburg, Jr. (1995). Where the Grass Grows Again: Knowledge Exchange in the Sustainable Agriculture Movement. *Rural Sociology* 60(4):721-740.

Herr, Andrew 2003. Industry Cluster Analysis of Westmoreland and Fayette Counties Addressing Workforce Development Needs. Prepared for the Westmoreland-Fayette Workforce Investment Board, October 14, 44 *pp*.

Hilchey, D. A Place at the Table: Explorations in Heritage Harvest Areas Development. In Denis Ebodaghe, Editor, *Proceedings of the Fourth National Small Farms Conference*, USDA, Washington, D.C., 2006, pp. 191-193.

Korsching, Peter F., and John C. Allen. (2004). Local Entrepreneurship: A Development Model Based On Community Interaction Field Theory. *Journal of the Community Development Society* 35:25-44.

Lee, Jo, Arnar Árnason, Andrea Nightingale and Mark Shucksmith. (2005). Networking: Social Capital and Identities in European Rural Development. *Sociologia Ruralis*, 45(4): 269-283.

Lobao, L. M. (1990). *Locality and Inequality: Farm and Industry Structure and Socioeconomic Conditions*. Albany, NY: State University of New York Press.

Marano, Cindy, Making Economic Development Ideas Work for Workers and the Economy: The Sector Approach. Keynote Address, Workshop for Planners in Workforce Development, Education, and Economic Development: Using Industry Cluster Analysis to Deploy Industry-Driven Programming, University Park, PA, June 17, 2004. www.nedlc.org/nnsp

NGA (2002). *A Governor's Guide to Cluster-Based Economic Development*, National Governor's Association, Washington, D.C., 47 *pp*.

O'Brian, David J., Andrew Raedeke, and Edward W. Hassinger. (1998). The social networks of leaders in more or less viable communities six years later: a reserach note. *Rural Sociology* 63(1): 109-126.

Porter, Michael E. (2000). Location, competition, and economic development: local clusters in a global economy. *Economic Development Quarterly*, 14(1): 15-34.

Porter, Michael E. (2004). *Competitiveness in Rural U.S. Regions: Learning and Research Agenda*, Harvard Business School Report, Institute for Strategy and Competitiveness, Feb., 70 *pp*.

Rogers, Everett M. (2003). *Diffusion of Innovations*. New York: Free Press. 5th ed.

Salamon, Sonya, Richard L. Farnswonh, and Jody A. Rendziak. 1998. Is Locally Led Conservation Planning Working? A Farm Town Case Study. *Rural Sociology* 63:214-234.

Scott, Allen J. (1992). The Collective Order of Flexible Production Agglomerations: Lessons for Local Economic Development Policy and Strategic Choice. *Economic Geography* 68:219-233.

Tolbert, Charles M., Thomas A. Lyson, and Michael D. Irwin. (1998). Local Capitalism, Civic Engagement, and Socioeconomic Well-being. *Social Forces* 77:401-428.

Warriner, G. Keith, and Trudy M. Moul. (1992). Kinship and Personal Communication Network Influences on the Adoption of Agriculture Conservation Technology. *Journal of Rural Studies* 8:279-291.

Wolf, Steven A., and Spencer D. Wood. (1997). Precision Farming: Environmental Legitimation, Commodification of Information, and Industrial Coordination. *Rural Sociology* 62(2):180-206.

This project is supported by the National Research Initiative of the USDA Cooperative State Research, Education and Extension Service, Small Farms and Rural Community Vitality Program, Grant No. 05-55618-15744.

Connecting Sustainable Agriculture to Rural Development:
The Case of Pasture-Based Dairy Grazing

Lois Wright Morton and Leah Miller

When economic rural development projects include a vision and goals for sustainable agricultural systems the capacity to achieve economic and social viability is increased. In this paper we offer a case study of how a group of Ohio dairy graziers have transformed their own and neighboring farms into economically successful pasture-based dairies and impacted farm communities in their region and beyond. Their success is a community development story that connects the building of technical and social infrastructures to create new rural economic opportunities. Some of the distinct characteristics of this grass roots ecosystem management model are citizen leadership, development of community through pasture walk groups, and the continual search and exchange of information and science to build the local knowledge base and capacity to adjust practices to localized conditions.

The community development "project" of renewing the vision and goals of rural regions across America to create economic health has often stalled because it has not simultaneously engaged the renewal of the agricultural sector. With increasing mainstream acceptance of sustainable agriculture, the opportunities for bridging community development and agriculture have increased. However, the processes needed to link the two are complex and based in the creation of multiple intersections among family farmers, existing and emerging science, practical technologies, local economies and governmental organizations (Green et al., 1993; Campbell, 1997; Lyson, 2004; Pigg and Crank, 2005). The transition to sustainable agricultural systems that renew the economic, environmental, and social conditions of rural places require two types of infrastructures: technical and social. The first, scientific and technical, is not just science and educational support but also access to financial and institutional resources. The second, social, is the building of public and personal connections to develop a community that facilitates farmer information exchanges, enables farmers to risk new ventures and adopt new technologies, provides political support needed for public understanding and acceptance, and links farmers to institutions, policies and resources. The community development practitioner is uniquely positioned to help the agricultural community make social and scientific connections among themselves and to the larger rural community. While many of the practitioner's favorite set of tools will work, the greatest challenge will be developing a vision of what is possible and transmitting it to others.

The infrastructure for sustainable rural places is being built place-by-place through a new kind of environmental movement called grass roots ecosystem management (Weber, 2000; Morton, forthcoming). The grass roots ecosystem management (GREM) movement "seeks to meld ecology with economics and the needs of community" through place-specific agency collaborations and citizen participation in rural economies that are dependent on their natural resource base (Weber, 2000, p.238). This movement is driven by the energies and local agricultural knowledge of citizens and the horizontal exchange of science and technological information that has practical applications to local problems.

In this paper we use the example of how a small group of Ohio dairy farmers have transformed their own and neighboring farms into economically successful pasture-based dairies and as a result have impacted farm communities in their region and beyond. Their success is a community development story that links the building of technical and social infrastructures to create new economic opportunities. The distinct characteristics of the dairy grazing movement include many grass roots ecosystem management strategies such as the development of community through pasture walk groups, emphasis on performance results (economic, environmental, social, and personal), and the continual search and exchange of information and science to build knowledge and capacity to adjust practices in their watershed to localized conditions.

We begin with the theory of grass roots ecosystem management (GREM) and its relationship to community development. Then a dairy graziers case study is offered as an example of how community social and economic infrastructures can be built to support transitions into sustainable agriculture. Lastly we end our paper with a discussion of the role of community development in linking sustainable agriculture to rural economies.

GRASS ROOTS ECOSYSTEM MANAGEMENT AS THEORY

Grass roots ecosystem management (GREM) is quietly emerging as an effective strategy for solving local and regional environmental problems (Weber, 2000; Morton, forthcoming). GREM utilizes a community development approach to managing local natural resources and defines community as "ecologically defined geographic space" (Weber, 2000, p. 247) rather than political or social boundaries. In this model place-based natural resources become the source of community identity as citizens find they share a common future when those resources are ignored, abused, or protected. It is the common identity and vision that unifies environmental, social, and economic agendas (Campbell, 1997). Abuse and careless management lead to degradation of the resource base which in turn reduce individual and community economic benefits, create conflicts among competing interests, and curtail the long term sustainability of place (Flora, 2001). "Conservation of natural capital creates economic value by maintaining the productivity of land and water resources while reducing 'externalities' such as ground water pollution or pesticide-related illness, whose costs are borne by the public" (Campbell, 1997, p. 38). Negotiation among the ecological community of landowners and citizens requires management of a local natural resource base so it is protected while being utilized for economic benefits. This management process cannot be solely accomplished by government directive and incentive programs but also requires the creation of social networks and relationship building among the diverse stakeholders within the community (Flora, 2002; 2004; Morton et al., 2006; Morton, forthcoming). Under this scenario, the community development goal is to connect the rural economy to the natural resource base in ways that engage citizens in public and private decisions that support humans *with* nature relationships rather than humans *over* nature.

The grass roots ecological management movement is characterized not by ideology but instead by a desire for concrete, on-the-ground results (Weber, 2000). Citizens and landowners who get involved want realistic, multi-functional solutions and use a number of strategies to build community, to manage conflict, to increase rural economic opportunities, and to protect their natural resource base. The initial strategy builds a social network by creating shared identity and connections to the local ecosystem. This social network is critical to the creation of developing a community of people who are not just personally invested in their natural resources but who are also concerned about their local economy. This is accomplished through community walking tours of the landscape, field inspections, first hand observance of how the ecosystem works, economic and social impacts and what is not working or could be improved. When these encounters with the landscape occur in group settings, citizens exchange personal knowledge and experiences that build agreement on environment issues and become catalysts for action.

Central to a local movement such as pasture-based dairy grazing is the production and transmission of knowledge through horizontal information exchange rather than formal and hierarchical structures of information distribution (Hassanein and Kloppenburg, Jr., 1995). GREM creates new knowledge as a social product through these horizontal interactions, cooperation, sharing and combining of information. Shared experiences frequently result in recognition of insufficient knowledge and lead to a search for more information (McCown, 2005; Morton, forthcoming). The community development practitioner's role becomes one of both process-building social connections and as resource to external science and technical services that can expand the local knowledge base in preparation for action. The goal is for science and technology transmissions to become continuous social transfers among engaged farmers, landowners and citizens and between local and expert knowledge sources.

One of the first group actions in connecting agriculture to the larger rural community is the collecting of data about the natural resource base by evaluating scientific reports related to local conditions, by compiling local knowledge and experiences, and by seeking technical advice and solutions. This is a critical point: as the community learns together, they begin to "own" the data (Epstein et al., 2006), sifting then to match their own ecosystem and goals for economic, social, and environmental outcomes. The continual community search, sort, and negotiation of multiple data are essential for achieving agreement and a willingness for individuals and the community to move to a "readiness" to experiment with solutions (Morton et al., 2006). Unique to this process is not a search for one "best" solution (Weber, 2000) but an understanding that science and knowledge continually change and that what works in one part of the ecosystem in a particular year or season may be less effective in another.

Sustainable agricultural systems have often incorporated modifications of the grass roots ecology management movement model in creating both a scientific and social support basis for the adoption of practices and systems thinking that are out of mainstream agriculture (Hassanein and Kloppenburg, Jr., 1995; Casler and Undersander, 2005). This "new" agriculture is actually a refinement of the pre-industrial/chemical-petro farming era with an emphasis on sustainable practices and family farms (National Catholic Rural Life Conference, 2007). The shift from confinement feeding systems to management intensive rotational grazing systems in the northern United States is an example of the new agricultural movement that began with livestock producer support groups that adapted routine methods and technologies to their personal farming values and needs through extensive exchanges among farmers (Hassanein and Kloppenburg, Jr., 1995; Casler and Undersander, 2005).

It is the dynamic efforts of community leaders that are committed to agriculture as a vocation and are willing to risk trying a new agriculture that holds promise for sustainable development and rural economic vitality (Lyson, 2004). The dairy grazing movement is one type of sustainable agriculture that has created a spill-over effect from individual farming decisions and practices to the larger social and ecological community. In the next sections we discuss dairy grazing agriculture and then offer a case study of an Ohio grazing community that grew from a small group of local farmers and is now transforming the way agriculture is perceived and connected to the rural region where they live.

CONFINEMENT DAIRIES VERSUS DAIRY GRAZING

Much has been written about the economic, social, and environmental sustainability of the confinement dairy, both supporting and challenging its basic premises (Dartt et al., 1999; Kriegl and McNair, 2005; Taylor and Foltz, 2006). Dairy farming in recent years has been an atomized occupation whose sole focus is to capture the economies of scale. With growth in herd size, these farms have developed elaborate feeding and bedding systems that demand complex manure handling systems (for removal and land application) in support of high density animal confinements (Dartt et al., 1999; Winsten et al., 2000a). Farmers have to grow, harvest and store feedstuff, including hay, haylage, corn silage, bedding, and corn year around or make off-farm purchases. The confinement dairy farmer needs both extensive labor and equipment. The cost of inputs frequently exceeds milk prices and the narrow margin of net gains and losses have led many farmers to leave the dairy business due to economic and personal time stresses of around-the-clock milking (Kriegl and McNair, 2005; Taylor and Foltz, 2006).

Dairy grazing is a counter movement to the "get big or get out" recommendation. These emerging alternative dairies combine a return to historical roots of dairying while applying new sciences of animal breeding, health and lactation, and manure and pasture management. Dairy grazing is a sustainable practice with strong positive environmental outcomes as well as economic and community impacts (Winsten et al., 2000a, 200b; White et al., 2002; Taylor and Foltz, 2006). Good pasture management practices utilize continuous cover crops (Hartwig and Ammon, 2002; Rochon et al., 2004), effectively distribute manure and its return as a nutrient to the soil (White et al., 2001; Cederberg and Mattsson, 2000), and aid herd health by reducing confinement and crowding conditions (Goldberg et al., 1992; White, 2000; Washburn et al., 2002). Conversely, poor pasture management results in ecological damage and animal nutritional deficiencies (Washburn et al., 2002), so technical and scientific knowledge exchanges are critical to the economic and environmental success of the farm operation.

Appropriate pasture management strategies can improve water quality by filtering nutrients while reducing runoff and high velocity flows of local streams (Hartwig and Ammon, 2002; Undersander et al., 2002; Gyssels and Poesen, 2003; Rochon et al., 2004). Moreover, well managed pastures provide habitat for diverse wildlife and visual green spaces in rural-urban interface zones (Rook and Tallowin, 2003; Radeloff et al., 2005). The economic benefits of grazing are low inputs and lower start up costs for beginning and transitional farmers (Winsten et al., 2000b; White et al., 2002). Dairy grazing also offers the marketing alternatives of organic and natural fed animal products (Cederberg and Mattsson, 2000; DeBoer, 2003). As one of the highest areas of return for farmers, dairy grazing can provide a sustainable and regular income for farm families (Dartt et al., 1999; Winsten et al., 2000b; Gloy et al., 2002; White et al., 2002) as well as quality time for personal lives and community activities (Undersander et al., 2002).

The profitability of dairy grazing finds its basis on reduced inputs, less labor and equipment and better animal health (Holmberg, 1992; Kriegl and McNair, 2005). Under this system, the cow hauls herself to the feedstuff in the field thereby reducing human labor and equipment inputs. Many dairy graziers do not raise grains, which can be labor intensive; they concentrate instead on growing grasses and legumes. Graziers will make the extra grasses and legumes into hay, baleage, or silage depending on their feed set up. But they are not under pressure everyday to be chopping or hauling feed. Confinement dairies can generate health issues related to the feet and transmission of diseases/illnesses that are not experienced as frequently by dairies that graze their cattle (Goldberg et. al, 1991; White, 2000; Kirkhorn and Garry, 2000; Washburn et al., 2002). Further, ventilation is an ongoing issue in confinement systems (Kirkhorn et al., 2001; Stowell et al., 2001; Wilkinson and Tilma, 2002; Gainet et al., 2007). All of these factors affect dairy cows, but they are limited in grazing systems because their confinement depends on weather during the winter. Thus labor, equipment, space, and lands are managed differently between graziers and conventional confinement dairies.

THE JOURNEY INTO DAIRY GRAZING

Methods

Documentation of this case study of the mid-Ohio dairy grazing experience is drawn from multiple sources over a seven year period. These sources include archival data from planning meetings for dairy conferences; public relations materials; conference agendas, proceedings, and evaluations; newspaper articles; personal correspondence; as well as interviews with key leaders in the community, including farmers, bankers, the Chamber of Commerce director, Ohio State University Cooperative Extension, Ohio Agricultural Research & Development Center (OARDC) scientists at Wooster and the United States Department of Agriculture/Agriculture Research Service/North Appalachian Experimental Watershed Station Coshocton Branch, director of The Small Farm Institute, and business leaders in the region. Farmers in the mid-Ohio region include Amish of German and Dutch descent and what is commonly called "English," meaning those who speak English and are not Amish. Both English and Amish farmers have crossed cultural barriers to create horizontal relationships around common needs for knowledge, and as a result have built a grazing community of farmers whose economic and social activities have spilled over into both English and Amish economies and social cultures.

Building the infrastructure

The transition to dairy grazing is not easy for confinement dairy farmers. Information and science are scarce and undervalued by many agricultural institutions as well as the farm community (Hassanein and Kloppenburg, Jr., 1995). Further, beginning farmers often experience financial and social pressures to conform to agriculture as currently practiced. One Amish dairy grazier comments on those who risk innovation: "Everyone is watching you to see if you are going to fail. As a new grazier, you need to do three things: pay your bills on time; keep your place neat; and never be late to church on Sunday. It helps to silence the critics" (Lloyd Miller, 2002). Thus, the development of a community network of graziers that can bolster each other by sharing technical knowledge and offering emotional support that the dairy grazing operation can be economically viable are critical. There are both technical and social infrastructure barriers to building a new agriculture that the mid-Ohio dairy grazers are working hard to transform.

TECHNICAL INFRASTRUCTURE

The goals of the dairy grazing movement are to be 1) more efficient in their use of natural resources (less fuel, fewer inputs); 2) better stewards of biological and natural resources (soil and water quality and plant diversity); 3) and effective rural entrepreneurs that support and strengthen their own finances and their local rural economy; while 4) providing quality of life for their families and community (Hassanein and Kloppenburg,

Jr., 1995; Dartt et al., 1999). The technical support mechanisms needed to accomplish these sustainable agriculture goals include access to relevant research and information, better communication and information flows from university scientists and experienced graziers, and access to financial support (Holmberg, 1992). The agricultural scientific community has been focused on the dairy confinement operation and often has no guidance to give (Hassanein and Kloppenburg, Jr., 1995; McNair, 2003). Much of the technical information is inaccessible (foreign sources – e.g. Fulkerson et al., 1997 and Cros et al., 2003 – or in out-of-print books), some of it is arcane, and some ecologically inappropriate. Many of the institutional structures to provide science and information are no longer in place (the Cooperative Extension Service is geared to advising large, full-time farms and confinement facilities) (Hassanein and Kloppenburg, Jr., 1995).

Moreover, the scientific information is a challenge since much of the grazing dairy research comes from Ireland, United Kingdom and New Zealand (Fulkerson et al., 1997) and needs to be adapted to local United States conditions. The limited grazing dairy research from the U.S. needs to be evaluated by region and scale. The other available information is from mentors and other graziers and is often anecdotal. The questions in the grazier's mind are: "How do I decide which practices are appropriate? Whose advice do I follow? What are the consequences of getting it wrong?" The grazier's goal is to minimize inputs and maximize outputs. They recognize that the outputs are greater than just the milk check. They look at the health of their livestock, the environmental impacts of their practices, the biodiversity of their pastures, the quality of family life and quantity of time for family and community.

The traditional technical and scientific infrastructure in agriculture is frequently formal, expert, and hierarchically filtered from institutional scientific research to farmers and focused on animal and individual farm level analysis and interventions (Hassanein and Kloppenburg, Jr., 1995) rather than targeting the whole community. Central to effective community development is the horizontal exchange of information and utilization of local knowledge on behalf of the community at large, not just the individual. However making use of what people know is messy and has its pitfalls (Innes, 1994; Pigg, 2005), and the farmer who chooses to move into sustainable agriculture is challenged to evaluate which information to put into practice and which to ignore. This is where a community of graziers can pool information and experiences and help each other by-pass the mistakes and the not-so-good results of others. The startup grazier can get into trouble following dairy advice from a conventional dairy farmer. The challenge for the new dairy grazier is to sort through the technical information offered and to decide what to try. One dairy grazier of 13 years says:

> I have often thought what I would do differently if I had to start again. First, ignore all the naysayers and those that said it could not be done. I listened to the wrong advice. The people that I sought out for advice were the established dairies who had farmed for generations. I was told to fix silos, grow corn, alfalfa, small grain, and keep the mows full of small bales of hay and straw also, keep a full line of equipment. The people of whom I had asked for advice really did not know what they were telling me was wrong for me. It was right for them, but… not for me. (Grim, 2006, p. 5)

The horizontal knowledge exchange of agricultural innovations in Appalachia and mid-Ohio was fostered by a regional not-for-profit organization founded in 2001 to respond to rural and agricultural issues to strengthen the rural economic base. This organization, Small Farm Institute (SFI), has a mission to work with small family farms and landowners of Ohio to help them increase farm profit options, to help farmers and rural communities with their environmental issues, and to create connections among rural communities and their local entrepreneurial farmers (http://www.smallfarminstitute.org/). Its funding is a partnership of public and private dollars including the Ohio Governor's Office of Appalachia, Ohio State University/Ohio Agricultural Research and Development Center and the United States Department of Agriculture/Agriculture Research Service/North Appalachian Experimental Watershed Station. The role of SFI is to connect farmers to each other, to scientific and financial experts, and to the larger rural economic development community. Increasing access to the latest dairy grazing practices and technical information using a community development strategy is one the institute's key activities. This community development process entails going beyond piping science from the land grant institution directly to the farmer and instead relies on the formation of information exchange networks among farmers; between farmers and their rural economic communities; and among the land grant institution, farmers, and the rural regions where they live and produce agricultural products. These networks are essential for bridging individual interests with the economic goals of the larger rural community.

A first community development effort was to connect scientists from the Ohio Agricultural Research and Development Center (OARDC) at Wooster and USDA Agricultural Research Service (ARS) Coshocton Branch to the grazier stakeholders' needs by meeting with researchers, writing joint grant proposals, and exposing the scientific community to the day-to-day challenges graziers experience. This activity started with the institute director and the chair of OARDC Agroecosystem Management Program (AMP) in 2002 and expanded to ARS Coshocton Branch in 2003 with the Seasonal Dairy Grazing Project. The goal: getting dairy grazing on the research agenda of the land grant university. The research needs span animal reproduction and lactation, nutrition and health; selection and growing grasses, legumes and pasture management for high milk production; as well as water quality-land use relationships. The second simultaneous goal was to create a community of graziers through conferences, field days, and pasture walk groups so they could exchange local and expert information and build new knowledge.

Prior to 2002, SFI was actively working with four dairy graziers who were seeking resources on how to manage their inputs to maximize their returns and profitability. Extension had gotten them started with workshops on managed intensive grazing (MIG) and dairy grazing. But they were hungry to know more and needed to connect with other dairy graziers. They wanted a conference to hear from other producers and other sources of information beyond the land grant universities and USDA ARS. These traditional research institutions did not have research initiatives in dairy grazing and were not aware of the growing interest in dairy grazing. SFI facilitated the first conference planned by local graziers in 2002 with 200 attendees; the 2006 conference drew over 500.

As the number of local dairy graziers grew, SFI encouraged farmers to interact more actively with local bankers, financial institutions, and chambers of commerce. Local and regional government leaders and economic development corporations have overlooked the possibilities in agriculture. At the same time, the agricultural sector has missed opportunities to interact because there has not been a common language or effective way to connect to local economic development and find common goals. Both have talked about quality of life and wanting to offer future generations opportunities in the community and the need to generate wealth (Wayne Economic Development Council Leadership Report, 2004). Initially, agriculture and rural entrepreneurs were not perceived as important to include in the race for rural economic revitalization. The focus was often on "high tech" jobs and "chain stores" instead of working and supporting the potential offered by building a diverse food supply system (Miller, 2006a). One of the main goals of the dairy graziers and SFI is to be seen as part of the economic future of rural communities.

The Director of Small Farm Institute writes:

> I have found that the local financial and political infrastructure will support larger scale agricultural operations because they understand the "monthly milk check" and "crop subsidies and risk insurance." However, these same financial and political institutions lack understanding and knowledge about similar benchmarks for a 'seasonal dairy,' an organic vegetable production, or a food processor working with local produce farmers to create "added value" to their surplus or grade produce. (Miller, 2006b)

A precursor to the formalization of SFI was a series of forums between farmers and several local bankers to help them learn from each other. On January 9, 2000, one of the first meetings was convened. Farmers attending made several points:

> Today's agriculture is multi-scaled and diverse, we need financial support... Further, large farms do not automatically make it a "factory farm" or corporation. It is producer philosophy and how the farm is operated that makes a difference among farms.... Suppliers are the creditors and lenders as a result of the 1980's credit crisis. When supplier credit has dried up and conventional credit is limited due to a lack of understanding of alternative agricultural farms, farmers who are good risks can't get financial support. (Meeting minutes)

Bankers responded that farmers need to know finances and how to run their farm as a business; farmers need to show their product, pay-back capacity and run the numbers. Indeed the banking community was not looking for agricultural loans and expressed the frustration that farmers cannot and often do not meet the standard credit factors the bank uses for loaning startup money or transitioning a business. The group has learned much from each other and made plans to continue to explore mutual benefits. They concluded that there is a need to connect to each other better to improve farmers' financial knowledge and how to grow the business, not just farm products. Further, they acknowledged these relationships were important to the overall economy of their region.

At a follow-up workshop in 2004, "Financial Lending Opportunities in the New Agriculture," farmers and bankers expanded their dialogue. Four farmers (produce, organic, grazing dairy, and wholesale/retail diary) introduced two concepts to the bankers: New Agriculture and benchmarking. The message to bank loan officers was that many of today's farmers are looking at products that could be labeled for environmental sustainability, such as *organic*. However, many banks are unfamiliar with the type of benchmarks that should be used to evaluate farm investment potential. Benchmarking is the identification of best practices that lead to superior performance of a business (Camp, 1989) and a process that industry has used for over 20 years. Farmers asked the bankers if they knew that an organic farmer had to be certified to sell organic products. They all said no. Farmers pointed out that the yield and practices would be different. Farmers asked the bankers if they could figure this out by looking at the financial statement of their operation. Again the answer was no. In 2004, consumer interest in local foods and organics had not yet hit the popular press (like it did in 2007). Many of the bankers were still thinking about 1990s agriculture. They did not know how a grazing dairy differed from a conventional, confinement dairy or how organic production differed from the industrial agricultural model they were used to financing. They didn't realize the seasonality of dairy grazing and that the farmer would not have income for two or three months, and loans needed to be designed to match income flow. They didn't know that calving would be in a short window in the spring and the milk volume would be greater in the spring grass flush than in the fall. Bankers needed new benchmarks that were different than traditional agriculture including knowledge about production practices as they related to organic and the certification.

This workshop became an example of what local communities could do to build a better connection among farmers and community organizations like chambers of commerce or economic development councils. Several local banks indicated that they had moved out of making agricultural loans. They did not see the future in agriculture. After the workshop, some bankers started attending the dairy conferences and field days as sponsors and vendors. They started advertising in local papers that they were offering agricultural loans. One loan officer that attended the workshop now takes fair week and does his business talking with farmers in the show ring, the sale barn and over French fries. He refuses to work the bank's fair booth because he's learned that the action is in the back. One banker mentioned after a 2007 field day event that it had taken over a year but he had three new customers because he developed a relationship with them (personal conversation with L. Miller, 2007). He said it would never have happened if he and the bank had not been introduced to these production practices.

SOCIAL INFRASTRUCTURE

The mid-Ohio dairy graizers are an informal diverse group with a large age range of young beginning farmers with families, middle aged part-time farmers who want to leave off-farm jobs to farm full-time, conventional confinement farmers thinking about transitioning to dairy grazing, and experienced graziers who have learned by the seat-of-their pants and welcome the chance to mentor and add to their knowledge (Demographic data from conferences, field days, and pasture walk groups, 2002-2006). The average age of dairy graziers attending conferences is under 40 years.

In addition to these demographic variations within the emerging grazing community, there are large cultural differences which include Amish and English cultures. The Amish community has a distinct, almost closed culture and network of trust that has often been wary of government programs (Kidder and Hostetler, 1990; Wasaoo and Donnermeyer, 1996; Wetmore, 2007). They are frequently not linked to agencies and organizations that service the agricultural community in their region (Conference evaluations, 2005). This means they may not understand the relationships between on-farm practices and water quality and need help in identifying sustainable agricultural practices that support environmental goals while not violating their cultural, community and religious beliefs. Through the development of the dairy grazing community, Amish and English are learning to trust each other because of their common farming identities and incrementally building social connections while exchanging technical and scientific knowledge.

In some ways, the Amish agricultural community is very much like the traditional English agricultural community in their approach to change. It is often slow and deliberative. They study and look at what others are doing before adopting change themselves (McCown, 2005). What is different is the strong group dynamics of the Amish community that provide either impetus or barriers to change. One long time Amish dairy grazier says that many young Amish men were approaching him about starting grazing dairies. The biggest obstacle young farmers face is the community. Fathers and grandfathers do not encourage them to farm, especially in the dairy industry. Thus, Amish wanting to adopt dairy grazing frequently experience strong community censure. These emerging leaders of the new agriculture need social networks from a community of dairy graziers in order to build confidence and have the capacity to risk change. These social networks are also sources of support for linking to technical and scientific information and community financial resources.

Leadership for the dairy initiative was drawn initially from a group of four graziers (English and Amish) who had attended general grazing conferences in Ohio and other states. Individually they were frustrated about the kind of information available to them, they were not specifically interested in organic and much of the dairy information was not relevant to their grazing management. In the SFI they found a resource that would help them articulate their vision for a different kind of dairying and help in building a community of graziers. As this group of committed graziers began to exchange their visions for dairy grazing with each other, they started the process of building a shared identity for themselves and others who might join the dairy grazing community. Although they had grazing interests in common, they were unlikely compatriots.

The first grazier is a former Rubbermaid marketing/sales person who travels with a calculator and makes most of his decisions based on return-on-investment learned from years at Rubbermaid. Originally a part-time farmer, in mid-life he jettisoned his off-farm job to farm full time. If the numbers do not show pay back for input, he figures out an alternative: share equipment, borrowe, lease or do without. He bought an old dairy farm in late 1990s of approximately 80 acres and currently milks 40 Jerseys with goals to manage the inputs, time for family and enjoyment of grazing cows along with protecting the environment with conservation practices. Located in the headwaters of the Killbuck Creek watershed, he participates in monitoring spring/tile drainage and samples water from his grazing pastures.

The second dairy graizer was raised on a dairy farm but his father had no farm to pass on so he bought an existing farm from a retiring dairy farmer. He had extensive problems with the cows and went to Wisconsin's dairy grazing programs to learn as much as he could about grazing. He has over 100 cows, involves his five children and wife actively in the farm operation and is excellent at networking with others. He has actively worked to get his Extension agricultural educator interested in dairy grazing along with efforts to get conventional confinement dairy farmer neighbors to switch to dairy grazing.

The third farmer is Amish and has been dairy grazing the family farm for the last 20 years. He is an innovative thinker who responds to ideas and is willing to experiment. Although the Amish community has frequently used him as an information source, he pushes the Amish community standards and practices with his early adoption and higher-than-usual willingness to take risks. This makes him an anomaly in the Amish community and he knows what it is like to experience community censure.

The fourth grazier is a horse and buggy Mennonite who moved from Lancaster, Penn., and an organic dairy grazier. He was one of the early individuals to be involved with a local group trying to manufacture and market organic dairy products. He has an extended family in his neighborhood that has also endorsed dairy grazing practices. This network of kin offer important learning exchanges and social support for his non-traditional dairy.

MERGING TECHNICAL AND SOCIAL

These four farmers were the catalyst for creating initial horizontal knowledge exchanges that ultimately led to a regional grazing movement. They decided to sponsor a conference so more local dairy farmers could learn about grazing. They gave themselves the name North Central Ohio Dairy Grazing Council (NCODGC) and began the process of doing something they had never done before. Through personal relationships with the director of SFI, they asked the institute to help them plan the conference and work through the logistics. SFI saw this as an agricultural economic development opportunity that needed facilitation and organization to succeed. The winter conference drew 200 people with the evaluation showing strong interest for additional grazing activities like pasture walks, tours and another conference. As a result, pasture walk groups were established in the spring of 2002. There are currently (2007) six pasture walk groups in the mid-Ohio region, each consisting of about 15 graziers who plan and convene monthly meetings. In the years following the first conference, sold out bus trips were offered to Wisconsin,

Pennsylvania, Virginia and Ohio for graziers to tour other pasture grazing farms. These successes reaffirm the importance of continuing to build the technical/scientific and social infrastructures if the graziers are to achieve their vision of community economic, social, and ecological sustainability.

The activities of the NCODGC in winter conferences, summer tours and pasture walks have led the SFI director to seek scientific and technical support for developing a farm management tool for decision-making (Klein, 1998; Paine, 2003). Farmer panels at the conferences frequently focus on decision-making. They ask, "What's the biggest mistake you have made in your grazing dairy operation?" or "What is your biggest success?" The majority of responses center on decision-making and the timing of the decision or that they recognize that they have to make a decision and there are alternative choices and need to evaluate the impacts of those choices.

Each year, new dairy graziers have joined the group; the 2006 dairy grazing conference had 500 participants attend over the two day conference. The original members of the NCODGC have rotated off and they have created a board that plans the conferences and other events. Thus, what began as a development process has been institutionalized, giving the movement sustainability beyond the passion and energy of its founders. The number of start-up graziers and beginners continues to grow. Breakout sessions for beginning and start-up graziers were requested for the 2006 conference. The conference evaluation reported 10 farmers who indicated that they would be starting to milk for the first time in 2006 and another 25 farmers who said that within the next two to three years they planned to start dairy grazing. Many of the new graziers have limited farming experiences, and if they do, it is not in dairy. In the Amish community, a number have fathers who work in wood shops or other manufacturing companies away from the farm, so they have limited farming opportunities. Some are leaving the shop or businesses to start farming. Another issue has been the negative response they have received from their church community in starting to farm. This has been a motivator to reach out to others in other church communities and locations in Ohio and across the United States.

A skeptical attitude about the future economic viability of farming occurs throughout the English and Amish farm communities in Holmes, Coshocton, Wayne and Ashland counties. The dairy grazing grass roots ecosystem management movement is countering this attitude and has made the region the center of dairy grazing for Ohio and the Amish-Mennonite community. Their social networks have extended beyond the region as different farmers from the North Central Ohio Dairy Grazing Council (NCODGC) have been invited to speak at conferences in other states (Wisconsin, Pennsylvania, Iowa, New York). As they have shared their experiences and the structure of NCODGC and their relationship with SFI, SFI has been asked how to form and maintain pasture walk groups and how to start and support new grazing dairy farms. For example, a group of Iowa farmers in 2006 asked for assistance in starting a dairy grazing pasture walk group in their community (Swackhamer et al., 2006). Some of the requests are because of the Amish network and the fact that all of the activities of SFI and NCODGC are farmer initiated, planned and implemented. These requests are not solely for individual farmer technical and scientific information but also for how to build the rural economic and social infrastructure needed to make local agriculture sustainable.

The grazing community has been building its own infrastructure through the development of leadership skills, mentoring, and identifying the decision making processes that young, new and more experienced dairy graziers need to be successful in agriculture. Some of the dairy graziers have started to transition to organics. Others have looked at developing added-value enterprises to their farms. Most are seeking ways to help their children be able to farm in the future. In addition, Holmes County Amish have started a Family Farm Field Day centered around grass based agriculture as a result of the activities of SFI and NCODGC. The first was in 2005 and over 1,800 people attended; and the second, held in July 2006, drew a similar number.

COMMUNITY DEVELOPMENT IN SUSTAINABLE AGRICULTURE

Community development has economic and social problem solving goals which utilize intervention strategies through local development, social action, or social planning modes (Rothman et al., 1995). Grass roots ecology management clearly falls into the local development model: practitioners act as coordinators, facilitators, and enabler-catalysts and create space for the community to be participants in an interactional problem-solving process (Rothman et al., 1995). For rural communities with a natural resource base such as agriculture, community developers must not overlook revitalizing agriculture as a component of economic development and community sustainability.

Sustainable agriculture is a type of farming that focuses on the system of humans *with* nature. While ecological systems are often the focal point, it would be a misunderstanding of sustainability if the vision does not include human social systems (Morton, 2003a; 2003b). The strength of sustainable agriculture and its diverse forms is its acknowledgement of the value of social relations and community. The grass roots ecological management model brings society – especially the community – into the decision making and management of natural resources. The mid-Ohio dairy grazing experience demonstrates how economic, environmental and social goals can be achieved in rural regions through processes that build the technical and social infrastructure.

Holmes County Chamber of Commerce Director Shasta Mast reiterates the connection:

> We must keep agriculture as part of our economic development strategy. The tourism segment of the Holmes County economy is dependent on having a rural landscape that contains working and profitable family farms. Further, the industry of cheese manufacturing and their milk demand has been one of the traditional economic resources. This needs to be expanded to build on consumer interest in local, seasonal, and fresh produce which has the potential to diversify our agricultural and tourism base. …It has been difficult to engage farmers in economic development because their farms are viewed as land bank for future development [house or commercial] and because no one has laid out how they might serve the community and what benefit there is to being involved in the local chamber of commerce or other economic development enterprises. (Personal Conversation, March 2005)

One role of the community developer is to lay out how agriculture and rural communities can serve each other. These graziers are transforming their regional agriculture and their rural economies by creating a new norm of acceptability for agriculture. Simultaneously, new leaders are being developed and are learning to articulate a shared vision for farming, land use, and rural economic sustainability. The dairy graziers have built confidence and feel empowered by their common vision to engage in mutual exchanges with researchers at Ohio State University/Ohio Agriculture Research and Development Center (OSU/OARDC) and USDA ARS North Appalachian Experimental Watershed in Coshocton. As a result of these relationships, they have become an important stakeholder base for these scientists. In addition, through the Small Farm Institute, dairy producers have connected with other producers in the Midwest and Northeast. These connections have broken their local isolation and built regional relationships with fellow graziers in other rural areas. As a result, a community of dairy graziers from rural places across Ohio and other neighboring states is being created. While this is a community of interest, its impact is on place-based communities. The application of community development processes has encouraged the replication of sustainable dairy practices and enabled these innovators to successfully challenge traditions. Further, they are no longer alone in their journey to establish grass-based dairies that neighbors and family members expected to fail. They have had to push against culture, community and family norms to change their production practices or start dairy grazing. But they have succeeded. One experienced dairy grazier sums up the individual and community impacts the grazing movement has had in the region:

> If I could have followed 13 years ago what I am advising today I know that our mortgage would be paid. What is the most exciting are all the success stories due to grazing....Today's startup grass based dairy has a tremendous amount of support that we did not have 13 years ago. Grass based dairying has evolved from a so-called fringe group, filled with kooks and nuts, and grown to a credible, dedicated, self-sustaining form of agriculture. (Grim, 2006)

What can a community development practitioner take away from this analysis of a grass roots-ecology management movement? First, practitioners need to help their community leaders recognize that agriculture has a future in rural communities and can be a source of economic growth. This means that economic developers need a vision of agriculture as part of the regional economic development plan. Second, the agricultural sector often needs the assistance of an intermediary, such as a community development specialist, to help them connect to each other and to the economic development community. Third, value locality development benefits when horizontal relationships are built and the local knowledge of the agricultural sector. Fourth, rural development success often begins with passionate local leaders who act as catalysts and partners with community development practitioners to create a sustainable infrastructure.

The community development process used with the dairy graziers is easily replicated to stimulate agricultural and natural resource based economies. These include a variation of the usual set of community developer tools: one-on-one exchanges that build trust; workshops for bankers with producers/agricultural enterprises; conferences that connect agricultural sectors to new agriculture possibilities; surveys of existing agriculture to discover what current economic needs are; small group meetings and dialogues among local farmers, businesses, and the chamber of commerce to develop vision and goals that include agriculture; and focus groups of consumers asking them what they would like from their local food system. While these are not new tools, their application to agriculture and natural resources is needed if rural places are to achieve their economic and social potential.

REFERENCES

Camp, R. C. (1989). *Benchmarking: The Search for Industry Best Practices that lead to Superior Performance.* Milwaukee, Wisconsin: ASQC Press.

Campbell, D. (1997). Community-controlled economic development as a strategic vision for the sustainable agriculture movement *American Journal of Alternative Agriculture.* 12 (1):37-44.

Casler, M. and D. Undersander (2005). "A bright future for grasses." Presentation given at the 2005 Wisconsin Fertilizer, Ag. Lime, and Pest Management Conference. 10 August, 2004. http://www.soils.wisc.edu/extension/wfapmc/2005/pap/Casler.pdf.

Cederberg, C. and B. Mattsson (2000). Life cycle assessment of milk production – a comparison of conventional and organic farming. *Journal of Cleaner Production.* Vol. 8, pp. 49-60.

Cros, M. J., Duru, M., Garcia, F. & Martin-Clouaire, R. (2003). Simulating Management Strategies: The Rotational Grazing Example, p.24, 32-34.

Dartt, B.A., J.W. Lloyd, B.R. Radke, J.R. Black, and J.B. Kaneene (1999). A comparison of profitability and economic efficiencies between management-intensive grazing and conventionally managed dairies in Michigan. *Journal of Dairy Science*. Vol. 82, pp. 2412-2420.

DeBoer, I.J.M. (2003). Environmental impact assessment of conventional and organic milk production. *Livestock Production Science*. Vol. 80, pp. 69-77.

Epstein, P. D., Coates, P. M., & Wray, L. D. (2006). *Results that Matter: Improving Communities by Engaging Citizens, Measuring Performance, and Getting Things Done.* San Francisco, CA: Jossey-Bass.

Flora, C. B. (2001). *Interactions Between Agroecosytems and Rural Communities*. New York: CRC Press.

Flora, C. B. (2002). Exploring the contents of the black box: rivers, people, and the places they live. NCRCRD, Iowa State University. Presentation at the EMAP Symposium May 9, Kansas City, MO.

Flora, C. B. (2004). Social aspects of small water systems. *Journal of Contemporary Water Research and Education* 128: 6-12.

Fulkerson, B., Blacklock, M. & Nelson, N. (1997). *Managing Pastures, Dairy Research and Development Corporation*. DairyLink, New Zealand.

Gainet, M, I. Thaon, V. Westeel, H. Chaudemanche, A.G. Venier, A. Dubiez, J.J. Laplante, and J-C. Dalphin (2007). Twelve-year longitudinal study of respiratory status in dairy farmers. *European Respiratory Journal*. Vol. 30, pp. 97-103.

Gloy, B.A., L.W. Tauer, and W. Knoblauch (2002). Profitability of grazing versus mechanical forage harvesting on New York dairy farms. *Journal of Dairy Science*. Vol. 85, pp. 2215-2222.

Goldberg, J.J., E.E. Wildman, J.W. Pankey, J.R. Kunkel, D.B. Howard, and B.M. Murphy (1992). The influence of intensively managed rotational grazing, traditional continuous grazing, and confinement housing on bulk tank milk quality and udder health. *Journal of Dairy Science*. Vol. 75, pp. 96-104.

Green, G. P., Flora, J. L., Flora, C. B. & Schmidt, F. E. (1993). From the Grassroots: Result of a National Study of Rural Self-Development Projects USDA ERS Agriculture and Rural Economy Division Staff Report N. AGES9325 Washington, DC.

Grim, E. (2006). The Grim Dairy. Proceedings of the North Central Ohio Grazing Conference for Dairy at Wooster/OARDC. January 26, 2006.

Gyssels, G. and J. Poesen (2003). The importance of plant root characteristics in controlling concentrated flow erosion rates. *Earth Surface Processes and Landforms*. Vol. 28, pp. 371-384.

Hartwig, N.L. and H.U. Ammon (2002). Cover crops and living mulches. *Weed Science*. Vol. 50, pp. 688-699.

Hassanein, N. and J. R. Kloppenburg, Jr. 1995. Where the grass grows again: Knowledge Exchange in the Sustainable Agriculture Movement. *Rural Sociology* 60:4:721-740.

Holmberg, J. (ed.) (1992). Making Development Sustainable: Redefining Institutions, Policy, and Economic. Washington, DC: Island Press.

Innes, J. E. (1994). *Knowledge and Public Policy*. 2nd Edition, New Brunswick: Transaction Publishers.

Kidder, R.L. and J.A. Hostetler (1990). Managing ideologies: harmony as ideology in Amish and Japanese societies. *Law & Society Review*. Vol. 24, No. 4, pp. 895-922.

Kirkhorn, S.R. and V.F. Garry (2000). Agricultural lung diseases. *Environmental Health Perspectives*. Vol. 108, Supplement 4: Occupational and Environmental Lund Diseases, pp. 705-712.

Klein, G. (1998). *Sources of Power How People Make Decisions* Cambridge: MIT Press.

Kriegl, T., and McNair, R. (2005). Pastures of plenty: Financial performance of Wisconsin grazing dairy farms. University of Wisconsin-Madison, February.

Lyson, T. A. (2004). *Civic Agriculture: Reconnecting Farm, Food and Community*. Medford, MA: Tufts University Press.

McCown, R. L. (2005). New Thinking About Farmer Decision Makers. In J. L. Hatfield (ed.), *The Farmer's Decision,* Chapter 2. Ankeny, IA: Soil and Water Conservation Society.

McNair, J. (ed.). (2003). Research aims to satisfy those who pay for it. *Graze Magazine*, March.

Miller, L. W. (2006a) Connecting Family Farmers and Economic Development. Kellogg Food, Society, and Agriculture essay. ---.

(2006b) Dairy Grazing: One Pathway to Sustainable Agriculture. Kellogg Food, Society,

and Agriculture essay.
Morton, L. W. (forthcoming). The Role of Civic Structure in Achieving Performance Based Watershed Management. *Society & Natural Resources*. ---.
(2003a). Civic Structure. In K. Christensen & D. Levinson (eds.), in *Encyclopedia of Community: From the Village to the Virtual World*. Thousand Oaks, CA: Sage. Pp. 179-182.---.
(2003b). Civic Watershed Communities. In M. M. Bell & F. T. Hendricks with A. Bacal (eds.), *Walking Towards Justice: Democratization in Rural Life Research in Rural Sociology and Development* Vol. 9 JAI/Elsevier, Amsterdam. Chapter 8, pp. 121-134.
Morton, L. W., Padgitt, S., Flora, J., Zacharakis-Jutz, J., Scholl, S., Jensen, A., Rodecap, J., Wagner, M., & Allen, B. (2006). *Renewing Local Watersheds: Leaders' Guide to Building Watershed Communities*. EPA CWA 10463 Project. Manual and CD ROM. Iowa State University. EDC 278 http://www.soc.iastate.edu/Extension/Watersheds_manual/index.htm
National Catholic Rural Life Conference. 2007. Creating a New Agriculture. http://www.ncrlc.com/Creating-New-Agriculture.html. Retrieved 8-22-2007.
Paine, L. (2003) The Art of Decision-Making October 2003 UW-Extension Columbia County.
Pigg, K. E. (2005). Introduction: Community informatics and community development. *Journal of Community Development Society,* 36(1): 1-8.
Pigg, K. E. & Crank, L. D. (2005). Do information communication technologies promote rural economic development? *Journal of Community Development Society*, 36:65-76.
Radeloff, V.C., R.B. Hammer, S.I. Stewart, J.S. Fried, S.S. Holcomb, and J.F. McKeefry (2005). The wildland-urban interface in the United States. *Ecological Applications*. Vol. 15, No. 3, pp. 799-805.
Rochon, J.J., C.J. Doyle, J.M. Greef, A. Hopkins, G. Molle, M. Sitzia, D. Scholefield, and C.J.
Smith (2004). Grazing legumes in Europe: a review of their status, management, benefits, research needs and future prospects. *Grass and Forage Science*. Vol. 59, pp. 197-214.
Rook, A.J. and J.R.B. Tallowin (2003). Grazing for pasture management for biodiversity benefit. *Animal Research*. Vol. 52, pp. 181-189.
Rothman, J. , J.L. Erlich, and J. E. Tropman. 1995. *Strategies of Community Intervention*. Fifth Edition. Itasca, Illinois: Peacock Publishers, Inc.
Stowell, R.R., C.A. Gooch, and S. Inglis (2001). Performance of tunnel ventilation for freestall dairy facilities as compared to natural ventilation with supplemental cooling fans. Livestock and Environment VI: Proceedings of the 6th International Symposium (21-23 May 2001, Louisville, Kentucky, USA), eds. Richard R. Stowell, Ray Bucklin and Robert W. Bottcher, pp. 29-40.
Swackhamer, C. with Miller, L. & Bloomfield Amish Dairy Graziers. (2006). Building Bloomfield Amish Dairy Graziers Pasture Walk Group Capacity. Proposal to Leopold Center for Sustainable Agriculture, Iowa competitive grants program.
Taylor, J. and Foltz, J. (2006). Grazing in the dairy state (Pasture use in the Wisconsin dairy industry, 1930-2003). University of Wisconsin-Madison (this one that financial and social (happy with lifestyle).
Undersander, D., Albert, B., Cosgrove, D., Johnson, D. & Peterson, P. (2002). Pastures for profit: A guide to rotational grazing. University of Wisconsin and University of Minnesota Extension Services.
Wasaoo, S.W. and J.F. Dennermeyer (1996). An analysis of factors related to parity among the Amish in northeast Ohio. *Population Studies*. Vol. 50, No. 2, pp. 235-246.
Washburn, S.P., S.L. White, J.T. Green, Jr., and G.A. Benson (2002). Reproduction, mastitis, and body condition of seasonally calved Holstein and Jersey cows in confinement or pasture systems. *Journal of Dairy Science*. Vol. 85, pp. 105-111.
Wayne Economic Development Council Leadership Report. (2004).
Weber, E. P. (2000). A new vanguard for the environment: Grass-roots ecosystem management as a new environmental movement. *Society & Natural Resources* 13: 237-259.
Wetmore, J.M. (2007). Amish technology: reinforcing values and building community. *IEEE Technology and Society Magazine*. Summer Issue, pp. 10-21.
White, S.L. (2000). Investigation of pasture and confinement dairy feeding systems using Jersey and Holstein cattle. Masters' Thesis, North Carolina State University.
White, S.L., R.E. Sheffield, S.P. Washburn, L.D. King, and J.T. Green, Jr. (2001). Spatial and time

distribution of dairy cattle excreta in an intensive pasture system. *Journal of Environmental Quality*. Vol. 30, pp. 2180-2187.

White, S.L., G.A. Benson, S.P. Washburn, and J.T. Green, Jr. (2002). Milk production and economic measures in confinement or pasture systems using seasonally calved Holstein and Jersey cows. *Journal of Dairy Science*. Vol. 85, pp. 95-104.

Wilkinson, R. and A. Tilma (2002). Livestock handling and confinement safety. Michigan State University Extension, National Ag. Safety Database. 10 August, 2007. http://www.cdc.gov/nasd/docs/d000701-d000800/d000754.html

Winsten, J.R., R.L. Parsons, and G.D. Hanson (2000a). Differentiated dairy grazing intensity in the Northeast. *Journal of Dairy Science*. Vol. 83, pp. 836-842.

Winsten, J.R., R.L. Parsons and G.D. Hanson (2000b). A profitability analysis of dairy feeding systems in the Northeast. *Agricultural and Resource Economics Review*. Vol. 29, No. 2, pp. 220-228.

Food System Makers: Motivational Frames for Catalyzing Agri-Food Development through Multi-Stakeholder Collaboration[1]

Wynne Wright, Michael Score, and David S. Conner

Multi-stakeholder partnerships are indispensable in the tool kit of community economic developers. In this paper, we examine the role of multi-stakeholder collaboration as a tool for renewing agriculture by using evidence from the Food System Economic Partnership (FSEP) in Michigan. FSEP represents a unique partnership of diverse - and in many cases, historically antagonistic - stakeholders from local government and universities, agri-business, and civil society. We present individual motivations of actors involved in this multi-stakeholder collaboration and their aspirations for agriculture as well as the collaborative experience. Using interview data from the partnership's leadership team, we examine the ways in which stakeholders frame their support for and involvement in food system collaboration. We identify two frames that explain motivation for participation in FSEP: agrarian fundamentalism and civic agriculture. These frames provide a look at how meaning is constructed around food system collaboration and informs future opportunities for sustainable community and economic development through cooperation. We conclude by identifying future directions for improvement of FSEP.

Agriculture long has been the economic and cultural mainstay in rural America. However, many communities are finding their markets less competitive, their social protections eroding, and their human capital deficient and outmoded and, thus, unable to meet the demands of the contemporary labor market (Beaulieu, 2002; Blakely and Bradshaw, 1985; Richardson, 2000). As a consequence, some are pooling their assets in creative ways to build robust community economic development initiatives through multi-stakeholder partnerships.

Multi-stakeholder partnerships are deemed indispensable in the tool kit of community economic developers (Logsdon, 1991; Pigg and Bradshaw, 2003; Savoie, 2000). It is neither sufficient nor efficient to rely on top down (McCall, 2003) or blue print planning (Bouwen and Taillieu, 2004) approaches to economic development. Rather, participatory engagement wherein people are seen as "agents rather than patients" (Sen, 2000, p. 137), and as actors capable of cooperatively addressing development concerns, now is an established practice, exclusion of which is deemed unacceptable and counterproductive.

Contemporary efforts at collaboration are opting for an integrative approach that invites and includes new partners. These new partners are treated as *active* (participating) members of the collaboration. Rather than merely being names on a partnership's membership roster to give the appearance of inclusivity, diversity, and collaboration, these new partnerships strive to be genuinely participatory. This includes incorporating those historically marginalized from development efforts. Pigg and Bradshaw (2003) argue that this form of empowerment – wherein citizens are given an authentic voice – is imperative to (catalytic) community development. The integration of new actors representing panoramically diverse and extensive sectors is in keeping with Granovetter's (1973) argument that development is enhanced when actors utilize "weak ties" which bridge untapped networks.

The body of literature focused on multi-stakeholder collaborations is rapidly growing in the fields of community economic development, public administration, and governance. An integrative model is increasingly prescribed for the revitalization of agriculture and food systems (hereafter, agri-food system) as well (Feenstra, 1997; Garrett and Feenstra, n.d., Winne et al., 1997). This trend corresponds to the mounting malaise of reductionistic science and its techno-fix orientation to agro-ecological problems, along with a desire for a course correction via the adoption of integrated food and farming systems' perspectives. Increasingly, food is being recognized as a lucrative domain for regional economic development (Friedmann, 2007; Lyson, 2004).

In this paper, the goal is to enhance our understanding of regional food system development by uncovering the motivations for multi-stakeholder collaboration using evidence from the Food System Economic Partnership (FSEP). FSEP is a five county economic development partnership in southeast Michigan devoted to catalyzing change in the region's agri-food system. Even though most of the participants are not in the position to benefit personally from any tangible outcome from FSEP, the leadership team is comprised largely of volunteers from local government and universities, agri-business, and civil society.

We present individual motivations of actors involved in this multi-stakeholder collaboration and their aspirations for agriculture and the collaborative experience. Using interviews from the partnership's board of directors and leadership team, we examine the social-psychological ways in which stakeholders frame (Benford and Snow, 1988; 2000) their support for and involvement in food system collaboration. Such frames are "critical to understanding how collaborative partnerships evolve and whether they will eventually succeed or fail" (Gray, 2004, p. 167).

Our interview data reveal two frames that explain stakeholder motivation for participation in FSEP: agrarian fundamentalism and civic agriculture. Rooted in neo-liberalism, agrarian fundamentalism makes no distinction between agriculture and rural communities, assuming what is good for agriculture is naturally good for rural communities and that benefits which accrue to agriculture will naturally spill over to communities. Civic agriculture, on the other hand, is a developmental approach to agriculture. It privileges the spatial embeddedness of production with special attention to the multiple benefits accruing from agriculture that both simultaneously benefit individuals and communities (Lyson, 2004). This paper reports the articulation of these two frames, based on interviews with the entire FSEP leadership team. We conclude with a reflexive look at how the organization can come closer to diffusing a community development approach to food system renewal. We begin by discussing the theoretical framework that informs our thesis.

Multi-Stakeholder Collaborations: Framing Food and Agriculture

Perhaps one of the most essential elements for reinvigorating community economic development is the adoption of a collaborative approach capable of mobilizing actors from a wide social, economic, and political spectrum (Botes and van Rensburg, 2000; Flora et al., 2001; Garrett and Feenstra, n.d.; Lowndes and Skelcher, 1998; Pigg and Bradshaw, 2003; Richardson, 2000; Winne et al., 1997). Collaboration is defined as "a process of facilitating and operating multiorganizational arrangements for solving problems that cannot be achieved, or achieved easily, by single organizations" (McGuire, 2000, p. 278).

This definition is sufficiently broad to create space for multiple stakeholders with differing worldviews to engage in the process of socially constructing meaning around the food system. It suggests evidence for recognition of the interdependence of social problems, thus requiring diverse talents, skills, and resources to address issues. The inclusive nature of multi-party collaboration explains, in part, the popularity of this community economic development strategy which has been marshaled to achieve a number of objectives from low-income housing (Bradshaw, 2000) to job training (Waddock, 1996) to environmental conflict (Everett and Jamal, 2004; Peterman, 2004). Lowndes and Skelcher (1998) contend that the field of urban and rural regeneration has been particularly fertile with multi-sector collaboration.

The reasons stated for the rise in multi-stakeholder collaboration are diverse. It is frequently argued that growth in collaborative endeavors is due in large part to a change or a decline in conventional institutional resources beginning in the 1970's (Bouwen and Taillieu, 2004; Lowndes and Skelcher, 1998). For example, Kettl (1996) argues that new technology and streams of knowledge have loosened the state's monopoly on information. This reshuffling of material and non-material resources has left the state searching for innovative ways to maintain needed public services (Lemann, 1994). Thus, collaboration has become a way to accomplish more with less. Partnerships add value "by bringing together complementary services and fostering innovation and synergy" as well as opening up new windows of opportunity that public or private agencies alone would be unable to achieve (Lowndes and Skelcher, 1998, p. 315). Bradshaw (2000) illustrates this point with the case of the U.S. Department of Agriculture's Fund for Rural America initiative which favors grant proposals that demonstrate an effort to pool resources as a condition of the award. Pooling resources in this way can also help partnerships become multi-functional as they serve several stakeholder needs (Bradshaw, 2000).

Others have argued that cross-stakeholder collaboration is an outgrowth of the mounting complexity of the issues facing government (Bradshaw, 2000; Rhodes, 1997; Stewart, 1996). It is also suggested that partnerships are a necessary part of the economic development engine as a way to subvert power relations that have traditionally marginalized some. Highly rationalized bureaucracies structured into hierarchal tiers have historically silenced the voices of select groups resulting in distrust, alienation, isolation and non-engagement (Putnam, 2000; Stoker, 1997; Wheeler, 1996). A growing recognition of regional, spatial, and cultural interconnectivity also has spurred partnerships (Flora, et al., 1997). Whether the connections be forged through resource amenities, ecosystems, or artisanal clusters, such "communities of interests [are] displacing the community of place" (Bradshaw, 2000, p. 137). Regional, as opposed to singular and insular community collaborations and partnerships, hold strong promise for realization of community development goals (Pigg and Bradshaw, 2003). Thus, a decided strength of FSEP is its regional rather than single community emphasis.

Huxham (1996) refers to the movement toward organized cooperation as the "collaborative advantage." The collaborative advantage comes in the form of "synergistic gain and programme enhancement from sharing resources, risks and rewards and the prioritizing of collaborative rather than competitive advantage" (Lowndes and Skelcher, 1998, p. 317). Winne et al. (1995; cited in Garrett and Feenstra, n.d., p. 6) describe a number of benefits that accrue from collaboration specifically for the agri-food system. These include the ability to tackle particularly complex issues by leveraging people, resources, and networks, the increased communication and 'buy-in' of local actors, and the enhanced opportunities for policy development among other benefits. Garrett and Feenstra (n.d., p. 7) conclude that, "[i]deally a coalition should include a balance of stakeholders within the local food system." But they do not go as far as to define what constitutes a stakeholder. We contend that any definition of a stakeholder in the food system must be sufficiently broad to include consumers and citizens who have long been overlooked as food system actors. Individuals must be encouraged to reclaim their citizen role and explore the impact of food system reorganization on community and social well-being (Wright and Middendorf, forthcoming).

Some have maintained that under the global restructuring of the economy and society, rural communities would do well to prepare to adapt to macro forces. This perspective denies communities their agency and assumes they are passive victims of broader economic forces. We contend that communities will be best served to withstand economic and social change by becoming proactive and preparing for vigilant engagement through multi-stakeholder collaboration. Following the lead of McCall (2003), this approach can allow communities to become *food system makers* rather than *food system takers* in the new global economy.[2]

In an effort to learn how individuals can position themselves to become 'food system makers' we analyze the motivations of FSEP members for organizing this economic partnership with the sociological concept of frame analysis. Initially identified in the work of Goffman (1974, p. 21), frames "locate, perceive, identify, and label" knowledge pertaining to the social world. Benford and Snow (2000, 1988) extend this perspective in their seminal work on collective action and social movements. For Benford and Snow (1992, p. 137), frames are interpretative schemata that are used to "simplify and condense the 'world out there' by selectively punctuating and encoding objects, situations, events, experiences and sequences of action within ones present or past environments." This conceptualization places the micro aspects of meaning – and the process of making meaning – at the center of analysis. Framing privileges a fluid and interactive construction of reality over a structural or determinist reading of social life. It can prevent analysts from "lapsing into either reification or treating meaning and ideas as given" or part of an inherent social structure detached from human agency (Mooney, 1990, p. 1).

Frames function as a repertoire of knowledge stored in memory. Like a tool in a carpenter's bag, frames, once constructed, can be retrieved to aid in organizing perception, providing guidelines for interpretation of social situations and for ensuing action (Johnston, 1995). In this way, frames facilitate sense-making, reduce ambiguity, align or diverge from existing frames within ourselves or among others, and guide action to reestablish order (DeWulf et al., 2004; Goffman, 1974). Frames "are not merely aggregations of individual attitudes and perceptions, but are also the outcome of negotiating shared meaning" (Gamson, 1992, p. 111).

Collective action, whether in the form of social movements or food system collaboration, requires the alignment of interpretative frames. Snow et al. (1986) refer to frame alignment as "the linkage of individual and SMO [social movement organization] interpretative orientations, such that some set of individual interests, values and beliefs, and SMO activities, goals, and ideology are congruent and complementary" (p.464). Although Snow et al., (1986) conceptually attribute frame alignment to social movement organizations, we find it illustrative to extend the concept to FSEP. We do not see FSEP as a social movement. FSEP is constituted by individuals who desire change in the food system, yet they go about this through relatively institutionalized means. The organization does not represent grievances about a particular set of issues nor organized frustration with the current food system. There is no evidence that members engage in symbolic politics to advance an agenda. On the contrary, most demonstrate a relative trust of economic arrangements and institutional politics. Many members are representatives of county government or the land grant university. Nonetheless, FSEP is a form of collective action and from an interactional perspective, we view it as a community development organization because it is intentionally and unintentionally engaged in "building relationships that increase the adaptive capacity of people who share a common territory" (Luloff and Bridger, 2003, p. 211). Thus, Snow and Benford's conceptual framework can be useful to the study of agri-food collective action and community development.

Snow and Benford (1988, 2000) have identified three functions played by frames. Diagnostic frames identify problems, label the situation as unjust, and attribute blame to an actor or structure. Prognostic framing advocates a line of action to remedy the condition. Lastly, motivational frames galvanize individuals to action by providing a rationale for redress. These three functions have been usefully applied to agriculture.

Mooney and Hunt (1996) found the framing concept constructive for identifying agrarian ideologies that constitute the fabric of social protest in American agriculture. They have documented three frames unifying agrarian social movements: agrarian fundamentalism, competitive capitalism, and producer frames (Mooney and Hunt, 1996). This study will reveal utility in the agrarian fundamentalism frame as it applies to FSEP collaboration. Agrarian fundamentalism elevates the role of agriculture in society and advocates for its restoration and protection, not for an ecological imperative or to nurture social relations but for the sake of supporting rural and urban economies.

Since the post Civil War era a dependency hypothesis has infused agrarian thought predicated on the idea that "the withdrawal of agricultural production … as an economic force … will make society suffer" (Mooney and Hunt, 1996, p. 185). This 'as goes agriculture, so goes the rest of the country' linear view underscores geo-political interconnections but locates agriculture at the base of all socio-economic life. The "specific form in which agriculture is organized is of less concern" than the advancement of agriculture itself (Mooney and Hunt, 1996, p. 183). This creates room for participants to tolerate a diverse structure of agriculture in regards to scale, ownership, or production practices. Hence, relatively little distinction is made between small or large producers and little distinction is made in regards to the social or ecological footprint of the farm enterprise. The agrarian fundamentalist frame is primarily driven by economics and social well-being springs from this foundation. We extend this work for the purpose of identifying the driving forces that propel actors to respond to a changing agri-food system through the vehicle of stakeholder collaboration.

In addition, we also discovered a second frame not originally articulated in the work of Mooney and Hunt (1996). We refer to this frame as "civic agriculture." Civic agriculture refers to the "emergence and growth of community-based agriculture and food production activities that not only meet consumer demands for fresh, safe, and locally produced foods, but create jobs, encourage entrepreneurship, and strengthen community identity" (Lyson, 2004, p. 2). Included under the rubric of civic agriculture are production strategies which increasingly are familiar to advocates of food system localization: community supported agriculture, community gardens, farmers' markets, roadside stands, u-pick farm patches, local processing, neighborhood kitchens, and locally controlled cooperatives. Civic agriculture serves multiple social needs, from community solidarity, integration, social equality, and democratic engagement to the need for individual profit. Unlike agrarian fundamentalism, civic agriculture exists between market and society.

We proceed by organizing this paper in two sections; we explore why multi-stakeholder collaboration has emerged as a tool for revitalizing the agri-food system in southeast Michigan. This necessitates a discussion of the larger social forces that constitute the socio-economic and political context facilitating collaboration. We then turn to the interview data which reveal the motivations of the FSEP participants, but first we introduce our case study.

Background: The Road to FSEP

The Food System Economic Partnership (FSEP) is a five county economic development food system collaborative in southeast Michigan. FSEP was officially launched in early 2005. It bills itself as a catalyst for change in the food system. More specifically, its mission is to strengthen the regional economy and preserve community character by innovating and expanding food system businesses and increasing consumer awareness of and access to local farm products in southeast Michigan. It does this by engaging in research, education, and outreach initiatives.

The FSEP concept came about through the merger of two relatively distinct initiatives. First, a group of graduate students at the University of Michigan (UM) jointly undertook a master's thesis project which designed a model for a multi-stakeholder collaboration to regionalize their agri-food system. Simultaneously, a Michigan State University Extension (MSUE) Educator began an informal dialogue with local community leaders to gauge their interest in collaborating with MSUE on how to respond to the economic downturn in Michigan agriculture and in the state generally. Ultimately, these two paths crossed, at which time it became clear that both saw agriculture as a way to re-invigorate community economic development. Students and UM faculty brought to the collaborative a conceptual platform on which to integrate stakeholders and begin the work of addressing the problems of agriculture and food in local communities. Extension educators contributed organizational acumen, an informed understanding of the biological and technical aspects of agriculture and the social fabric of rural Michigan, as well as extensive networks of potential partners.

The students and extension educator began by giving presentations to county commissioners and other local authorities in the five counties that make up their region. This includes the counties of Wayne, Washtenaw, Jackson, Lenawee, and Monroe. Collectively, these counties cover 3,332 square miles, including the Detroit metropolitan area, and adjacent, primarily small townships and rural areas. They solicited support from local county governments for the organization and maintenance of a regional food system partnership. Each participating county government contributed $15,000 and continues this investment annually.

Structurally, FSEP is comprised of a leadership team and a board of directors. The leadership team consists of 22 members and is made up of farmers, agri-food distributors, economic development planners, agricultural lenders, public heath and nutrition specialists, locally owned restaurant personnel, county commissioners, Michigan State University Specialists and Extension staff, and representatives from the Michigan Coalition of Black Farmers, Michigan Farmers Union, Organic Growers of Michigan, Michigan Organic Food and Farming Alliance, and Slow Food Huron Valley. The role of the leadership team is to create and implement a strategic action plan aligned with the FSEP mission.

The board of directors is made up of seven voting members. It includes one representative from each of the five participating counties and two at-large members elected by the leadership team. The board is charged with control and management of business, property, personnel, and operating funds. After one year, FSEP applied for and received 501C3 (non-profit) status. For the past two years the collaborative has been funded through a combination of grants and county government annual contributions. FSEP has also been awarded a number of modest grants totaling $50,000.

Early in the organizational life of FSEP, considerable investment was made in capacity building. This took the form of crafting a shared vision, articulating organizational goals, and identifying skills for leadership. FSEP also embarked upon a strategy key to capacity building, which consisted of enlisting a diverse leadership team in order that local resources could be marshaled to catalyze regional development. This strategy has been employed successfully in numerous projects, one of which is the project known as EPIC – Environmental Programs/Partnerships in Communities (Richardson, 2000). Pigg and Bradshaw's theory of catalytic community development (Pigg and Bradshaw, 2003) identifies such capacity building as foundational to effective and sustainable community development.

Through this process, FSEP participants adopted six goals:

1. Improve the viability of the agricultural sector in the region by identifying consumer demand and helping farm and food system businesses and entrepreneurs convert that demand into new ventures (e.g., processing facilities, marketing campaigns, value chain partnerships, new infrastructure).
2. Provide revitalization opportunities for metropolitan Detroit and other urban areas in the region through business development and job creation, increase demand and incentives for the re-use and redevelopment of older buildings, and reduce chronic unemployment.
3. Create wealth for the region as an array of diverse new businesses and economic development ventures are launched and revenue growth is strategically reinvested by project partners in additional ventures and infrastructure.
4. Improve consumer understanding of what is produced, processed and marketed in the region.
5. Improve producer understanding of consumer needs.
6. Improve land use policy and decision-making over the long-term by translating demand side knowledge of the food system into economic opportunities.

These goals suggest that FSEP, as an organization, has taken a neo-liberal or market-based approach to community economic development. Evidence from the interpretative frames articulated by members, however, will show that this orientation is not shared by all.

Methods

The analysis presented here relies on data derived from 22 structured interviews in 2006. This includes interviews with the entire population of FSEP participants; however, one of the respondents was not actively involved in FSEP at the time of the interview.[3] No distinctions were made between leadership team members and board members; our goal was to avoid taking a sample of participants and, given the small number of actors involved in this initiative, a census was reasonable. Interviews typically lasted between 45 minutes to 1 hour and took place in the respondent's home, place of work, or neighborhood coffee shop.

After inquiring about the respondents' programmatic relationship with FSEP (e.g., length of involvement, roles, linkages with other sectors), we asked them to reflect on why they were involved with FSEP, the impact of FSEP on their worldview, what accomplishments they hoped FSEP might achieve, their perceptions of the partnership's strengths, as well as barriers to FSEP's work. For the purposes of this paper, we are particularly interested in exploring motivations offered by respondents to extend our understanding of food system collaboration.

FSEP members span the market, state, and civil society gamut. Although members represent both rural and urban areas, most of the respondents had an urban background. This composition is deliberate. New participants are recruited to the project by leadership team members through invitation, taking care to ensure the broadest representation as possible. This membership composition underscores the increasing participation of non-farm actors in the food system along with the growing movement to reengage the agri-food system as a vital economic development stream for both rural and urban areas.

The Work of FSEP

FSEP has engaged in a number of projects to invigorate economic growth in southeast Michigan. Objectives in the partnership are accomplished through working groups organized around five general thematic areas. Table 1 provides a snapshot of the priorities of each working group along with a description of their mission.

Table 1.

Working Group	Objectives and Sample Projects
Farm-to-School	The farm-to-school project is designed to increase the use of local produce in public school meal programs. Three pilot school lunch programs will be kicked off in 2007.
Business Innovation and Networking	This work group assists entrepreneurs with development of business concepts and written business plans. FSEP has worked on business plan development with more than 20 entrepreneurial teams and is assisting in a feasibility study for bio-diesel production.
Research and Software Development	Initial research has focused on the flow of agricultural commodities from farm to retail outlet. Plans continue for research to define gaps in the food economy and identify how regional or national economic trends impact Michigan's potential for business growth. They also partner with the MSU Product Center for Agriculture and Natural Resources to develop software tools for entrepreneurial business planning.
Education and Outreach	As progress is made in research or project development, FSEP works through its Education and Outreach team to disseminate findings. Education is conducted through a series of quarterly newsletters, website, and through an annual educational conference. The first conference in 2006 was attended by more than 150 people.
Membership	The goal is to maintain a diverse and integrated team of stakeholders from the regional food system including those who have historically held oppositional views. The membership group conducts an on-going assessment of leadership team composition to maintain the desired balance.

The FSEP Context

New partnerships for community development are embedded in an expansive social, economic, and political context, inclusive of institutions, norms, and values, which can either hinder or lend a hand toward realizing stakeholder goals. We begin this analysis of FSEP by exploring these contextual forces that facilitated the conditions for multi-stakeholder collaboration. They are: 1) demographic shifts, 2) depressed business climate, 3) changes in land use policy, 4) agri-food globalization, and 5) pre-existing models of development, specifically the operation of a new ethanol production facility in the area.

Demographic Shifts

Much of Michigan continues to be very rural, but the five-county FSEP area boasts 27 percent of the entire state's population. A good deal of the target area has experienced population growth over the past 15 years, but more recently, the region as a whole is starting to experience net out-migration above the national average. Census data show that there has been a decline in the overall population of the region; between 2000 and 2003, southeast Michigan lost 9,000 people per year (SEMCOG, 2004). Much can be attributed to a reshuffling of residency patterns within the region.

The metropolitan centerpiece of southeast Michigan is the urban hub of Detroit, located in Wayne County. The city and surrounding metropolitan area is home to over two million people and ranked tenth nationally in terms of population and twelfth in terms of employment. In recent years, however, Detroit and Wayne County have experienced a steady stream of out-migration as residents resettled in neighboring counties. Those departing tend to be young (20-34), nearing retirement or retired (over 50), and on average, those who are better educated, adding to the difficulties for economic renewal (SEMCOG, 2004).

Depressed Business Climate

Those exiting are leaving behind a city facing a future of persistent job loss and mired in unemployment. Much of this mobility is due to a depressed manufacturing economy that has been plagued by inability to provide good jobs (SEMCOG, 2004). The Detroit metropolitan statistical area boasts one of the highest unemployment rates in the nation – 6.7 percent (U.S. Dept. of Labor, 2006). Wayne County lost 76,000 jobs in the past five years and is expected to lose between 3,500 and 4,000 in the coming years (University of Michigan, 2006). In 2004, almost 19 percent of individuals in Wayne County had incomes below the poverty line (U.S. Dept of Agriculture, ERS, 2004).

Economic analyses of the region paint a grim picture and although business fluctuation persists, data reveal a struggling state tied to the boom and bust cycles of a monoeconomy. Economists project that putting so many of its economic eggs in the automobile industry basket has left Michigan lagging behind the nation in economic performance. A stagnating economy, fractured labor force, dwindling tax base, and loss of human capital make economic innovation emergent. While problems abound, Detroit and southeast Michigan are rich in other assets which can be recombined in new ways to create opportunity. One of those roots is diversified land use conventions that can help pave the way to economic renewal.

Favorable Land Use Policy

The history of land use and economic development policy in Michigan is, for the most part, a local history. The state pioneered land use and zoning policy by permitting decision-making around land use to occur at the local level. Public Act 207 (1921) permitted zoning in all cities and villages. Soon after, the act was followed by passage of township and county zoning acts. Planning acts passed between 1931 and 1945 authorized regional and county level planning. These original acts remain in place, largely unchanged, resulting in a system that authorizes the smallest jurisdictions, which have adopted planning zoning ordinances, to determine their own development destinies. Likewise, the 1940s through the 1980s witnessed a great deal of legislation aimed at economic and urban development which paved the way for local governments to address the specific needs of individual communities (Michigan Land Use Leadership Council, 2003).

This grassroots authority enables local governments to address pressing issues without coordinating their actions with local planning commissions or through the burdensome creation of new ordinances. Local governments are allowed to respond to public health, safety, or public welfare issues without specific enabling legislation as long as state laws are not violated. In essence these land use policies signal the presence of political opportunity for food system collaboration. Political opportunity refers to an absence of institutional barriers inside the state that might circumvent any form of political mobilization in society (McAdam, 1982). In short, a political opportunity exists when cooperative and consistent policy or agency support aligns with the goals of those organizing for change. Michigan's openness to diverse models of land use and the consistent deployment of those guidelines suggest that appeals for local economic development are more readily received by governmental actors and state resources are made available to realize these goals. This helps explain the 'buy-in' of local governments to contribute financial resources and supply volunteers to serve on FSEP's leadership team.

Agri-Food Globalization

The recent trajectory of Michigan's agricultural production sector – including the FSEP area - mirrors nationwide trends. Farms are declining and this fate has particularly hit hard mid-sized operations. Overall, Michigan experienced a loss of at least 21 percent of its farms between 1974 and 2002, while the land in farming decreased by only about 7 percent, implying consolidation of farmland into larger units. Earning profit in agriculture has become illusive - only about 43 percent of Michigan farms reported positive net earnings in 2002 (National Agricultural Statistics Service, 2002). The reasons for these trends are numerous and complex, but can be explained, in part, by the general trend toward the globalization of the agri-food system.

The global restructuring of the agri-food system refers to the reorganization of national economies growing out of *Pax Americana* (McMichael, 1996). In this process, national regulatory mechanisms gave way to socio-economic and political realignment (Friedmann, 1991; McMichael, 1994) aimed at the "functional integration between internationally dispersed activities" (Dicken, 1992, p.1). This agri-food globalization has deep roots in mercantilism but intensified in the latter quarter of the twentieth century. For agriculture, this process refers to the hyper-mobility of trans-national capital and the consonant deregulation of agricultural trade policy that has, in effect, restructured national agri-food systems into global supply chains. The implications of agri-food globalization vary by academic discipline. A number of economists (Blank, 2001; Rickard, 2004), for example, see the loss of farm numbers and land devoted to agriculture in a generally positive light, resulting in a more efficient agricultural system and lower food costs for consumers: "Thus, painful as the process must be for individual farmers, overall the Schumpeterian process of creative destruction ensures that productive resources …are transferred to more productive uses, and it is this process that underpins rising living standards" (Rickard, 2004, p.749). Technology and knowledge, not people, are seen as drivers of this process and largely treated as exogenous, even inevitable forces of change (Blank, 2001; Rickard, 2004).

In contrast, other social scientists trained to identify the non-economic implications of global restructuring see the process introducing new vulnerabilities for social relations and ecosystem health. Some of these vulnerabilities include the withering of our domestic food base, increased food insecurity, decreased plant biodiversity, increased food miles, community transitions, a reshuffling of national, state and local political power, and community fragmentation (Koc and Dahlberg, 1999; Phillips, 2006;). The process of globalization is described as resulting in increased concentration of ownership and control of the agri-food system (and therefore power imbalances) by trans-national corporations, achieved through horizontal and vertical integration and global sourcing (Heffernan, 2000). In these accounts, the process is facilitated by public policy and research that elevates excess food production while decreasing labor use, along with the deregulation of markets and securities (Bonanno and Constance, 2003; Heffernan, 2000; Heffernan and Constance, 1994; Koc and Dahlberg, 1999).

The global restructuring of the agri-food sector is not a *fait accompli*: it is being contested on a number of fronts (Whatmore and Thorne, 1997). The political and socio-cultural nature of these contests present themselves in a diversity of ways from campaigns to ban genetically modified organisms, to the protests against global trade, to a rise of a new entrepreneurial ethos in producers and the increasing engagement of consumers. For example, growing consumer preference for locally grown and processed food is well documented (see, e.g., Feenstra, 1997; Ostrom, 2006; Vander Mey, 2004), despite some racial differences (Zepeda and Leviten-Reid, 2004) and confusion about what constitutes "local food" (Ostrom, 2006). Both those who advocate for the global mobility of people and capital and those who view it as a "race to the bottom" agree that the ultimate losers are smaller scale farmers producing undifferentiated bulk commodities for mass markets. FSEP can, therefore, be seen as an effort to resist this process by crafting new markets and differentiating products which emphasize their local and regional nature, while at the same time responding to consumers' preferences for locally grown foods. Said another way, FSEP can help foster community development and serve as a hedge against "defensive localism" (Flora, 2001; Hinrichs, 2003), which can seriously hamper community vitality and viability.

Pre-existing Models of Development: Lenawee Ethanol Production Facility

Collaboration between MSUE, Ohio State University, and corn and soybean marketing programs in the two states resulted in the successful conception and execution of an ethanol production facility expected to convert more than 18 million bushels of grain annually into fuel grade ethanol, distiller's grain, and commercial grade carbon dioxide. The material result was a stimulus to regional grain markets, but on a cognitive level this successful venture inspired optimism for new agriculture development. The enterprise not only convinced farm leaders that change was possible but taught them if left uncoordinated, change in the food system might undermine their investment in the ethanol market. Because products from the ethanol plant were to be sold largely into regional urban markets, leaders appreciated the need to promote agri-food system planning that involved urban and rural communities. Mutual benefits to urban and rural interests seemed important in promoting an integrated development initiative that would appeal to leaders in public and private sectors.

The above discussion is meant to provide context for understanding the formation of this multi-stakeholder partnership for regional food system development in Michigan. As we have seen, it is in this region of shifting demographics, in a state where local governments can use a combination of planning/zoning ordinances with the authority to act independently within broader state laws to address public needs, that FSEP was formed. It is also a setting of fluctuating economic activity in general, and a global restructuring of agriculture in particular, that set the stage for FSEP. Lastly, as the discussion above suggests, the formation of FSEP was also stimulated by the concrete success achieved in developing an ethanol production facility in the region. These five socio-economic and political dynamics constituted the structural conditions in which individuals could orchestrate a regional partnership and become 'food system makers'. However, even when conducive structural conditions are present it remains necessary for individuals to engage in the interpretative work needed to catalyze collaboration.

McAdam (1982) has found that when conditions such as "recession," "loss," or "opportunity" are diagnosed, a mental transformation, or "cognitive liberation" may take place causing individuals to make a value judgment about the state of events. If new frames are constructed to define the situation as "problematic," "unsustainable," or "fortuitous," the seeds for action may be sown. Snow and Benford (1992) demonstrate in their work on the nuclear freeze movement that "frame resonance" is pivotal to successful collective action. In the section below we turn to the interview data which underscore the role of frame resonance to impede or facilitate food system collaboration.

Framing Agri-Food System Collaboration

We use the framing perspective to help us identify the interpretative work in which FSEP members engage. Interpretative frames amplify "existing beliefs and values, most of which are associated with existing ideologies" that operate outside the individual but can be drawn upon to advance collective action goals (Snow and Benford, 2005, p. 209). Our analysis of the motivations described by the 22 participants revealed two interpretative frames: agrarian fundamentalism and civic agriculture. These frames bring people together to jointly engage in the work of food system development. Below we discuss the ways in which these frames were articulated allowing us to trace "the subjective meanings people attach to their objective situations" (McAdam, 1982, p. 34).

Agrarian Fundamentalism Frame

The agrarian fundamentalism frame locates agriculture as a preserve of economic opportunity and posits the dependence of urban and rural communities on agriculture. Some might refer to this as the 'old agricultural paradigm' because of the exaltation of agriculture. Agrarian fundamentalism tolerates a diverse structure of agriculture in regards to scale, ownership, and production practices to provide the food for a subsistence handicapped urban and rural non-farm population. This interpretative frame was shared by 60 percent of FSEP members. Perhaps the best example typifying the agrarian fundamentalism frame is offered by the following respondent:

> The value of agriculture in society has to be the highest, but the pay is the lowest of all. We need to care about where food comes from and reward those people... We have a diverse group of people on the [FSEP] team, but some members seem to be shortsighted – they don't see the value in agriculture, especially if it is not food or organic. They push a focus on food too much rather than what's good for the agricultural economy.

The thread running through much of the agrarian fundamentalism frames was the homogeneity of agriculture. All agriculture was deemed virtually the same, thus eclipsing the fragmentation that separates community members and denying the challenges faced by farm families (Hinrichs, 2003). The treatment of agriculture as a homogenous form of production also can be seen among those who were passionate about agriculture as an instrument for economic development. One member states, "agriculture has become isolated from mainstream economic development...we need to help people see the importance of agriculture and its ability to shape the economy in rural areas but the importance of agriculture for urban areas too."

Not only was the primacy of agriculture stressed, but issues like scale and ownership structure were less essential to the agrarian fundamentalism frame. This group was more likely to view any agricultural development as positive and to be less critical of scale, ownership structures, labor practices, or production practices (e.g., organic vs. non-organic). One member of the leadership team bemoaned her frustration with FSEP saying, "there was not anyone from the large chain stores represented on the team – like [regional supermarket chains such as] Meijers, Spartans, or Wal-Mart." This individual was a conventional farmer. Most of the respondents who framed their membership in FSEP as a function of agrarian fundamentalism were affiliated with occupations or organizations that would typically be thought of as 'conventional'.

Agrarian fundamentalism was most frequently articulated in the form of encouraging a 'value-added' agriculture by a majority of the agrarian fundamentalism framers. Indeed, 80 percent of the agrarian fundamentalist framers specifically indicated that they joined FSEP to "add value to agriculture" through activities seemingly as diverse as developing ethanol production facilities or creating niche markets for grass fed beef. A number of members commented that one of the most promising ways for FSEP to achieve its goals was to realize a success story – some tangible project capable of showcasing agricultural entrepreneurialism through the development of local markets. Almost all pointed to the prospect of the Farm-to-School Project which brings fresh food to school cafeterias as holding the most potential for a veritable success story. Allen and Guthman (2006) argue, however, that Farm-to School programs "have for the most part accepted neoliberal notions of governance such as individualism, entrepreneurship, choice, and devolution" thus reinforcing the ascendancy of agriculture markets as opposed to crafting meaningful alternatives that allow for the questioning of existing economic arrangements (p. 412). The neo-liberal orientation is congruent with the agrarian fundamentalism frame.

In recounting their motivations, a holistic view of 'the food system' receded somewhat and conventional agricultural production, processing, and distribution forces were given prominence in the frame. Consumers were articulated as partners for what they could inform producers about their desires and consumption habits, not as actors responsible for social, ecological, or economic sustainability. Place was integrated into the frame for its ability to supply unique resources that might underpin market niches or reduce input costs, not as a unique geography which roots people "in a common purpose and for nurturing a sense of belonging to a place and an organic sense of citizenship" (DeLind, 2002, p. 217). It is also notable that neither those experiencing food insecurity, hunger nor their representatives are engaged in the partnership. Environmental concerns as they interface with the food system are not on the radar of members either; only one respondent regarded ecological concerns as a motive for instituting change in the food system. Collectively, these characteristics reinforce the collaborative as an exercise in agricultural productivity.

This effort to close the boundaries around the agri-food system and exalt agricultural productivity and marketness (Block, 1990) above social functions excludes new actors from food system integration and can promote divisiveness. It valorizes agriculture as an uncritical production form and silences any discussion of the multiple functions a sustainable agriculture system can play in society beyond subsistence, such as nurturing community social life, supporting human health, and balancing ecosystems. We see this as a form of instrumental cooperation that relies on collaboration as a last gasp for survival yet retains boundaries and suspicion. This instrumental approach is akin to defensive food system localization (Hinrichs, 2003).

Those adhering to an agrarian fundamentalism frame showed little interest in discussing how to go about resolving these differences or what form of organization could help achieve this goal. They preferred to focus on 'doing' the work of adding value to agriculture. One leadership team member summed up what a majority of members articulated as a frustration of the collaboration:

> We spent a lot of time floundering at the first. We spent too much time thinking about who we were and engaged in capacity building and not enough time 'doing' ag. marketing. A lot of the producers left in the early stage because they had the attitude 'call us when you are ready to do something.'

Another member echoed this tension between attention to capacity building and market development:

> I joined at the beginning and was religious about it for a time, but we spent so much time devoted to strategic planning that we did not get to the heart of business innovation and networking...My time as an entrepreneur is lost when I am navel gazing. I spend 80 percent of my time on business and 20 percent on my organization but FSEP spends 80 percent on soft stuff and 20 percent on building networks for business.

Agrarian fundamentalists did not challenge the political economy of agriculture; instead they confined their interest to developing new markets that could compete on a regional or national level. One member summarized this orientation when she stated,

> Two groups are emerging in FSEP: those who understand the economics of commodity agriculture and have made our peace with it and those who are process oriented and focus on inclusivity. The result is a real separation of ideas where people get alienated.

Such group divisiveness was a common theme running through most interviews. These comments reflect the tensions that arise when differing frames collide.

Civic Agriculture Frame

Forty percent of the respondents in this study attributed their motivation for food system collaboration to a multidimensional frame we have labeled "civic agriculture." Most of the respondents using this frame were affiliated with occupations or institutions having little or no connection to conventional agriculture. For the most part, these respondents represented non-profit organizations.

Civic agriculture is frequently articulated as a perspective that privileges civic engagement and smaller locally-based enterprises. Proponents extol its multi-functional character by combining profit-making with other societal needs, and in this way, departing from the economically rational actor that occupies a more central position in the market dogma of agrarian fundamentalism. Thus, civic agriculture recognizes the multiplicity of roles individuals play (e.g., producer, eater, neighbor, parent, teacher, churchgoer, friend, etc.). This multi-functional orientation can serve as an engine for economic development and community building simultaneously. This frame is validated through social science empirical contributions of Goldschmidt (1978) and Lyson (2004). The civic agriculture frame was employed by one leadership team member who stated,

> Many economic actions are based in social needs. For example, the farm economy is a means to achieve farmland preservation among other things. If we take an economic development approach we will get caught up in the same old system. This would not help people achieve their goals for long term sustainability, public health, and access to local food. If we [FSEP] are to be a catalyst of change we have to build community, forge relationships, and build an appreciation for the diversity of people, ideas, and ways of knowing the world.

Another member reflects the difficulty in preserving extra-economic values as they pursue a locally embedded food system when she states,

> We need to facilitate a market distribution system but our challenge is, how do we make sure the business works to meet the needs of a local food economy? How do we retain non-economic values in these start-up ventures we nurture? They need to maintain our mission and vision of a sustainable food system.

At the heart of this frame is a spatial dimension that embeds food production in locality and in this way privileges smaller scale enterprises, unlike the agrarian fundamentalism frame. As one FSEP member explained, "economic viability depends on a regional approach to the food system. That is our key mission and we [FSEP] have to focus on [developing] mechanisms for community agriculture or short supply chains."

Relationships also become centralized in the civic agriculture frame, as illustrated by the following statement: "I saw FSEP as a way to start healing problems. People need to start to talk to each other. I saw this group as a way to start a dialogue and create social equity among various actors in the food system." Talk of relationship building, social equity, and extra-economic social problems were common themes running throughout the frames articulated by those proclaiming civic agriculture. Agrarian fundamentalism is based on a 'rising tide' thesis that assumes economic gains accruing to agriculture will overflow in abundance and somehow magically find its way into community development coffers. But this assumes an objective character to community development, such as infrastructural development via the building of roads, schools, water treatment plants, job creation, business expansion or other efforts to 'grow' physically and financially. Wilkinson (1991) referred to this notion of community as development in community.

Civic agriculture belongs to the development of community (Wilkinson, 1991) school of thought and is rooted in an interactionist approach. It assumes routes to profit-making and community-building can be taken through multiple approaches (e.g., community-supported agriculture, farmers' markets, community processing facilities) but the key is economic arrangements which are embedded in community. Community integrated enterprises promote economic development but at the same time force social interaction due to the limited scale on which they operate. They force people to come together more regularly – they structure in opportunities for visibility and interaction rather than anonymity and formality. Locally based enterprises allow individuals to collaborate to solve shared problems and in this way, relationship building becomes central to community development. From an interactional approach, the mere act of working cooperatively is sufficient to begin building relationships and nurturing community solidarity (Wilkinson, 1991). "Communities develop and well-being is enhanced when residents work together to address common issues and problems – that is, when they exhibit community agency" (Luloff and Bridger, 2003, p. 212). The following quote is from a member who stated that she joined FSEP because she was attracted to just such an interactionist view of agriculture and rural development:

> I was interested in different dimensions of the local food economy such as rural revitalization, farm and open space preservation, reducing food miles. Agriculture is where humanity meets ecology and when we make the connection visible we can create a socially, ecologically, and economically sustainable food system.

Also in contrast to agrarian fundamentalism, civic agriculture is less restricted in regards to membership. The participation of actors "across different dimensions" of society is encouraged. As one member noted when reflecting on her aspirations for FSEP:

> Our goal is to attempt to be as inclusive as possible. You should build on your assets and be people-based. Build on people in your community who have interest in this sort of project and invite your enemies too. Think outside your box and invite people who don't share your views so you ensure diversity, motivation, and interest.

Another member shared this view by advocating inclusivity:

> Pitch a big tent and focus on shared values. Forget that he is a democrat or republican or that she is an economic developer or a 'foodie'. We all share a vision of a viable community and local food system.

Membership diversity has presented conflict in the form of tension between strict entrepreneurial activity (those eager to get on with project development, also called the "do-ers") and those who saw process ("the processors") as essential. At other times, respondents used the labels "strategic planners" versus "the applied mechanists" or the "crunchies" versus the "aggies." It should come as no surprise that those framing their involvement with an agrarian fundamentalist frame predominately overlap with the "aggie" label and the "civic agriculture" framers are aligned with the "crunchies" for their focus on process, relationships, and extra-economic outcomes.

This counter-trend agriculture is framed as offering an integrative vision promoting local environmental stewardship in its sensitivity to place-based biological and cultural needs over the reductionism of conventional agriculture. Consumers find their demand for fresh, high quality produce satisfied while rural community culture is nurtured through enhanced food system literacy and social interaction (cf., Feenstra, 1997; Groh and McFadden, 2000). These social functions are accomplished while simultaneously creating engines for economic development for farmers, local artisans, and others with a creative entrepreneurial vision. In this way, it liberates communities, food systems, and people from global dependence by promoting paths of local interdependence. We suspect its timely emergence is due to a loss of agriculture's "experiential commensurability", or resonance with participants' worldview, and the rise of a growing population free of any ties to agriculture production, yet who understand the role of food in society through a culture of consumption.

Discussion and Conclusion

Cooperation has become fundamental to agriculture development for over a century, but more recently it has taken a different turn. Boundaries around who is considered an appropriate cooperator are being loosened and diversity, not homogeneity, is becoming a defining element. As a result, the agri-food system has become more accessible to a broader population. Commodity organizations, marketing boards, and other commodity specific industries and associations which dominated the structure of agriculture markets and politics for much of the twentieth century, are now finding they have to share the agri-food organizational terrain with multi-stakeholder organizations whose strength lies in public engagement rather than technical expertise or economic analysis.

Our research has probed the motivational frames of collaborators for food system development. Our discussion of civic agriculture and agrarian fundamentalist frames attempted to clarify the motivations of participants by proposing two frames that catalyze participation. The two groups differed on their prognosis of the problem – those with an agrarian fundamentalism frame advocated more market based approaches to agriculture development whereas the civic agriculture framers contended that agriculture should be developed by social processes like civic engagement and community as priorities. This perspective rests on the premise that localities that embrace such an orientation will experience multiple benefits from economic stimulation to community well-being. These two frames are fundamentally different (Lyson and Tolbert, 2003). Given the differences, it is useful at this point to consider how these frames originally intersected.

We attribute the presence of these two frames to the original organizational structure adopted by the early organizers of FSEP. Specific organizations and individuals were extended an invitation to participate in the FSEP leadership team by the extension educator and organizing students. The networks of the extension educator were pivotal as this individual straddles differing ideological terrains in his region. His extension connections in the land grant put him squarely in the domain of conventional agriculture, yet he has worked hard to reach out to others historically neglected by the land grant university system and to build networks with those working for a regional agri-food system. Nonetheless, membership in the early days was based on who the extension educator and students invited. The top-down nature of the early organization certainly challenges respondents' claims of diversity, inclusivity, and participation.

Although most respondents spoke of the importance of diversity and participation these claims do not appear self-critical when stacked against the evidence. As noted by other respondents, there are no representatives from the supermarket sector, there are no voices from the world of food security, nor is there input from traditional commodity organizations. *It is imperative to note, however, that commodity organizations have been* invited to participate and have declined. Moreover, the role of the membership committee is to regularly scrutinize the membership composition to evaluate their shortcomings and needs for new members. The reality is, however, as all community organizers realize, that such collaboration cannot be forced. You can, however, build such a dynamic organization that others see your organization as *the* partner to have and recognize the merits of collaboration. This has yet to occur for FSEP. Regardless of these unfulfilled goals, this organizational structure has resulted in a sufficiently diverse membership body who articulated the dualistic interpretative frames of civic agriculture and agrarian fundamentalism. As membership continues to grow and become more diverse, we can expect to see a greater degree of diversity in the frames of the participants. This will increase the need for FSEP to negotiate frame contests. One way to avert dysfunctional contests is to highlight the points where frames align.

Both groups share an appreciation of the role of agriculture in society and agriculture's ability to address certain social needs whether it be purely for economic development as with the agrarian fundamentalists or to address multidimensional needs as with the civic agriculture frame. The key, however, is in their joint interest – collaborating for agri-food renewal and, in the process of collaboration, community is being developed.

Secondly, both groups shared the rhetoric of sustainability, even though they went about operationalizing this concept in different ways. The agrarian fundamentalists tended to adopt a hierarchal approach to sustainability viewing it through an economic development lens. They privileged opportunities to generate profit, believing that social and ecological needs would spring from this success. The civic agriculture framers were more likely to view sustainability in a more multi-dimensional way, seeing economics as no more or less integral than ecological integrity or social relations.

Lastly, we have demonstrated that multi-stakeholder collaboration is accompanied by different frames that can be aligned with or in opposition to one another. Nonetheless, collaboration can proceed apart from the competing frames. Hence, many respondents indicated that inter-group tension was a central ingredient in FSEP, allowing diverse groups to be integrated into the work of sustainability, yet consistent throughout all frames was an appreciation for diversity, if they have yet to fully achieve this goal. To quote one FSEP participant, "our strength lies in our diversity. That is the source of our strength and our weakness." Another member remarked, "The unique and great thing about FSEP is bringing a variety of people together who would not normally sit together … Having [those who disagree with you] on the board is necessary but it makes progress rocky." Collectively, these frames provide the rationale for participating in this voluntary multi-stakeholder coalition but at the same time infuse the friction to promote organizational dynamism. This dynamism is necessary to propel the organization forward. Retaining oppositional stances is critical within a context of collaboration.

FSEP is designed to be – and should be – a self-correcting enterprise. In order to achieve its mission of being an effective and valuable tool for community development, FSEP must be evaluated to ensure that it is adhering to its basic paradigmatic premises (e.g., ensuring full and diverse representation) and is fostering measurable outcomes for the communities engaged in and being served by FSEP.

It is not clear as to whether FSEP will establish itself as an important player in the field of community economic development or catalyze change in the food system as their organizational directives mandate. The organization must learn to push the boundaries of inclusivity and engagement while managing the competing frames that result from such diversity. This enhances reflexivity and illuminates the ways in which food and agriculture weave us all into a larger social tapestry, and as such, oblige us to pursue food system citizenship.

NOTES

1 An earlier version of this paper was presented at the 2007 Agriculture and Human Values Society annual meetings in Victoria, British Columbia. We wish to thank the editors and two anonymous reviewers for their helpful comments on previous drafts of this manuscript.

2 This framing is built upon the work of McCall (2003, p. 7) who advises governments to adopt supportive rather than directive approaches to community economic development so that communities can become "development makers rather than development takers."

3 This participant had only recently ceased participation in the leadership team of FSEP. His remarks to the interviewer suggest he had grown frustrated with the organization's inability to rapidly create new markets for farmers. I have not elaborated on the institutional affiliation of the respondents and their corresponding frames for fear of revealing their identity. Given the small number of respondents in this study and the fact that there often are only one or two people from each sector, it would too easily be possible to identify a respondent's response if information regarding their corresponding frame was provided.

REFERENCES

Beaulieu, L.J. (2002). *Creating vibrant communities and economies in rural America.* Southern Rural Development Center, Publication No. 225. Mississippi State, MS: Southern Rural Development Center. Retrieved August 16, 2007 from http://srdc.msstate.edu.

Benford, R. and Snow, D. (2000). Framing processes and social movements: An overview and assessment. *Annual Review of Sociology,* 26, 611-39.

Botes, L. and van Rensburg, D. (2000). Community participation in development: Nine plagues and twelve commandments. *Community Development Journal* 35(1), 41-58.

Blakely, E. J. and Bradshaw, T.K. (1985). Rural America: The community development frontier. In F. A. Fear and H.K. Schwarzeller (Eds.), *Focus on community: Research in rural sociology and rural development* (Vol. 3, pp. 3-29). Greenwich, CT: JAI Press.

Blank, S.C. (2001). Globalization, cropping choices, and profitability in American agriculture. *Journal of Agricultural & Applied Economics*, 33(2), 315-326.

Block, F. (1990). *Postindustrial possibilities: A critique of economic discourse.* Berkeley: University of California Press.

Bonanno, A. and Constance, D.H. (2003). The global/local interface. In D. L. Brown and L. E. Swanson (Eds.), *Challenges for rural America in the twenty-first century* (pp. 241-251). University Park, PA: Pennsylvania State University Press.

Bouwen, R. and Tailler, T. (2004). Multi-party collaboration as social leaning for interdependence: Developing relational knowing for sustainable natural resource management. *Journal of Community and Applied Social Psychology*, 14(3), 137-53.

Bradshaw, T. K. (2000). Complex community development projects: collaboration, comprehensive programs, and community coalitions in complex society. *Community Development Journal,* 35(2), 133-145.

Butler, L.M. and Carkner, R. (2001). Bridges to sustainability: Links between agriculture, community, and ecosystems. In C. B. Flora (Ed.), *Interactions between agroecosystems and rural communities* (pp. 157-174). Boca Raton, FL: CRC Press.

Campbell, D. (1997). Community-controlled economic development as a strategic vision for the sustainable agriculture movement. *American Journal of Alternative Agriculture,* 12(1), 37-46.

DeLind, L.B. (2002). Place, work and civic agriculture: Common fields for cultivation.*Agriculture and Human Values*, 19(3), 217-224.

DeWulf, A, Craps, M. and Dercon, G. (2004). How issues get framed and reframed when different communities meet: A multi-level analysis of a collaborative soil conservation initiative in the Ecuadorian Andes. *Journal of Community and Applied Social Psychology,* 14, 177-192.

Dicken, P. (1992). *Global Shift.* (2nd ed.). New York: Guilford Press.

Everett, J and Jamal, T.B. (2004). Multistakeholder collaboration as symbolic marketplace and pedagogic practice. *Journal of Management Inquiry*, 33(1), 57-78.

Feenstra, G. W. (1997). Local food systems and sustainable communities. *American Journal of Sustainable Agriculture,* 12(1), 28-36.

Flora, C. B. (2001). Shifting agroecosystems and communities. In C. B. Flora (Ed.), *Interactions between agroecosystems and rural communities* (pp. 5-13). Boca Raton, FL: CRC Press.

Flora, C. B. and Flora, J.L. (2003). Social capital. In D. L. Brown and L. E. Swanson (Eds.), *Challenges for rural America in the twenty-first century* (pp. 214-227). University Park, PA: Pennsylvania State University Press.

Flora, J. L., Sharp, J., Flora, C B. and Newton, B. (1997). Entrepreneurial social infrastructure and locally-initiated economic development in the nonmetropolitan United States. *Sociological Quarterly* 38(4), 623-45.

Friedmann, H. (1991). Changes in the international division of labor: Agri-food complexes and export agriculture. In W.H. Friedland, L. Busch, F. Buttel, and A. Rudy. (Eds.), *Towards a new political economy of agriculture* (pp. 65-93). Boulder, CO: Westview.---.

(2007). Scaling up: Bringing public institutions and food service corporations into the project of a local, sustainable food system in Ontario. *Agriculture and Human Values,* 24(3), 389-398.

Gamson, W. A. (1992). *Talking politics.* New York: Cambridge University Press.

Garrett, S. and Feenstra, G. (n.d.) *Growing a community food system.* Retrieved on December 19, 2006 from http://smallfarms.wsu.edu/publications/WREP0135.pdf.

Goffman, E. (1974). *Frame analysis: An essay on the organization of the experience.* New York: Harper Colophon.

Goldschmidt, W. (1978). *As you sow: Three studies in the social consequences of agribusiness.* Montclair, N.J.: Allanhold, Osmun and Co.

Granovetter, M. (1973). The strength of weak ties. *American Journal of Sociology*, 78(6), 1360-1380.

Groh, T. and McFadden, S. (2000). *Farms of tomorrow revisited: Community supported farms farm supported communities.* San Francisco, CA: BIO-DYNAMIC Farming and Gardening Association, Inc.

Gray, B. (1989). *Collaborating: Finding common ground for multiparty problems.* San Francisco: Josey-Bass.---.

(2004). Strong opposition: Frame-based resistance to collaboration. *Journal of Community and Applied Psychology*, 14 (3), 166-176.

Heffernan, W.D. (2000). Concentration of ownership and control in agriculture. In F. Magdoff, J.B. Foster, F. Buttel (Eds.), *Hungry for profit. The agribusiness threat to farmers, food, and the environment* (pp. 61-75). New York: Monthly Review Press.

Heffernan, W.D. and Constance, D. H. (1994). Transnational corporations and the globalization of the food system. In A. Bonanno, L. Busch, W. Friedland, L. Gouveia, and E. Mingione (Eds.), *From Columbus to ConAgra: The globalization of agriculture and food* (pp. 29-51). Lawrence, KS: University of Kansas Press.

Hinrichs, C. C. (2003). The practice and politics of food system localization. *Journal of Rural Studies,* 19(1), 33-45.

Huxham, C. (1996). *Creating collaborative advantage.* London: Sage.

Johnston, H. (1995). A methodology for frame analysis: From discourse to cognitive schemata. In H. Johnston and B. Klandermans (Eds.), *Social movements and culture* (pp. 217-246). Minneapolis: University of Minnesota Press.

Kettl, D. F. (1996). Governing at the millennium. In J.L. Perry (Ed.), *Handbook of public administration* (2nd ed., pp. 3-18). San Francisco: Josey-Bass.

Koc, M. and Dahlberg, K. (1999). The restructuring of food systems: Trends, research and policy issues. *Agriculture and Human Values* 16, 109-116.

Lemann, N. (1994). The myth of community development. *New York Times Magazine*, January.

Logsdon, J. (1991). Interests and interdependence in the formation of social problem-solving collaborations. *Journal of Applied Behavioral Science,* 27(1), 23-37.

Lowndes, V. and Skelcher, C. (1998). The dynamics of multi-organizational partnerships: An analysis of changing modes of governance. *Public Administration*, 76, 313-33.

Luloff, A.E. and Bridger, J.C. (2003). Community agency and local Development. In D. L. Brown and L. E. Swanson (Eds.), *Challenges for rural America in the twenty-first century* (Pp. 203-213). University Park, PA: Pennsylvania State University Press.

Lyson, T. A. (2000). Moving toward civic agriculture. *Choices* (Third Quarter), 42-45.---. (2004). *Civic agriculture: Reconnecting farm, food, and community*. Tufts University.

Lyson, T.A. and Tolbert, C. M. (2003). Civil society, civic communities, and rural development. In D. L. Brown and L. E. Swanson (Eds.), *Challenges for rural America in the twenty-first century* (pp. 228-238). University Park, PA: Pennsylvania State University Press.

McAdam, D. (1982). *Political process and the development of black insurgency, 1930-1970.* Chicago: University of Chicago Press.

McCall, T. (2003). Institutional design for community economic development models: Issues of opportunity and capacity. *Community Development Journal,* 38(2), 96-108.

McGuire, M. (2000). Collaborative policy making and administration: The operational demands of local economic development. *Economic Development Quarterly* 14(3), 276-291.

McMichael, P. (1994). The global restructuring of agro-food systems. Ithaca, NY: Cornell University Press. ---.

(1996). *Development and social change: A global perspective.* Thousand Oaks, CA: Pine Forge Press

Michigan Land Use Leadership Council. (2003). A brief history of land use and related policy in Michigan. Lansing, MI: Michigan Land Use Leadership Council. Retrieved 4, 2007, from http://www.michiganlanduse.org/resources/councilresources.htm.

Mooney, P. H. (1990). The ideological constitution of agrarian social movements in the United States. Paper Presented at the Annual Meeting of the European Society of Rural Sociology, Giessen, Germany.

Mooney, P. H. and Hunt, S.A. (1996). A repertoire of interpretations: Master frames and ideological continuity in U.S. agrarian mobilization. *Sociological Quarterly*, 37(1), 177-197.

National Agricultural Statistics Service. (2002). Census of agriculture. Retrieved January 5, 2007 from http://www.nass.usda.gov/Census_of_Agriculture/index.asp

Ostrom, M. (2006). Everyday meanings of 'local food:' Views from home and field. *Community Development*, 37(1), 65-78.

Peterman, W. (2004). Advocacy vs. collaboration: Comparing inclusionary community planning models. *Community Development Journal*, 39(3), 266-276.

Phillips, L. (2006). Food and globalization. *Annual Review of Sociology*, 35, 35-57.

Pigg, K. and Bradshaw, T. (2003). Catalytic community development: A theory of practice for changing rural society. In D. L. Brown and L. E. Swanson (Eds.), *Challenges for rural America in the twenty-first century* (pp. 385-396). University Park, PA: Pennsylvania State University Press.

Putnam, R. (2000). *Bowling alone: The collapse and revival of American community*. New York: Simon & Schuster.

Rhoades, R. (1997). Forward. In W. Kickert, E. H. Klijn and J. Koppenjan (Eds.) *Managing complex networks: Strategies for the public sector* (pp. xi-xvii). London: Sage.

Richardson, J. (2000). *Partnerships in communities: Reweaving the fabric of rural America.* Washington, DC: Island Press.

Rickard, S. (2004). CAP reform, competitiveness and sustainability. *Journal of the Science of Food and Agriculture,* 84, 745-756.

Savoie, D.J. (2000). *Community economic development in Atlantic Canada: False hope or panacea?* Canadian Institute for Research on Regional Development: University of Moncton, Moncton, CA.

SEMCOG, The Southeast Michigan Council of Governments. (2004). Migration and its impacton southeast Michigan, 1990-2003. Southeast Michigan Council of Governments, Detroit, MI.

Sen, A. (2000). *Development as freedom*. New York: Anchor Books.

Snow, D. and Benford, R. (1988). Ideology, frame resonance, and participant mobilization. *International Social Movement Research*, 1, 197-218.---.

(1992). Master frames and cycles of protest. In A. D. Morris and C. McClug Mueller (Eds.), *Frontiers in social movement theory* (pp. 133-55). New Haven: Yale University

Press.---.
(2005). Clarifying the relationship between framing and ideology. In H. Johnston and J. A. Noakes (Eds.), Frames of protest: Social movements and the framing perspective (pp. 205-11). Lanham, MD: Rowman and Littlefield.

Stewart, J. (1995). A future for local authorities as community government. In J. Stewart and G. Stoker (Eds.), *Local Government in the 1990s.* London: Macmillan.

Stoker, G. (1997). Local political participation. In R. Hambleton and Associates, *New perspectives on local governance.* York: York Publishing Services.

United States Dept. of Labor, Bureau of Labor Statistics. (2006). Metropolitan area employment and unemployment. Washington, DC. U.S. Dept. of Labor, Retrieved January 5, 2007 from http://www.bls.gov/schedule/archives/metro_nr.htm#BM2004.

United States Dept. of Agriculture, Economic Research Service (ERS). (2004). 2004 County-level poverty rates for Michigan. Washington, DC., U.S. Dept. of Agriculture, ERS. Retrieved January7,2007,from http://www.ers.usda.gov/Data/PovertyRates/PovListpct.asp?ST=MI&view=Percent.

University of Michigan. 2006. Wayne county job losses will be moderate through 2008. Ann Arbor, MI: University of Michigan. Retrieved January 12, 2007, from http://www.ns.umich.edu/htdocs/releases/print.php?Releases/2006/Apr06/r040506.

Vander Mey, B. J. (2004). The globalization of food and how Americans feel about it: Results of two surveys. *Journal of Food Distribution Research,* 35(1), 6-17.

Waddock, S. (1986). Public-private partnership as social product and process. In J. Post (Ed.), *Research in corporate performance and policy* (Vol 8, pp. 273-300). Greenwich, CT, JAI.

Weingartner, P. and Mooney, P. H. (1994). The ideology of sustainability: Frame alignment, master frames, and collective action frames and the sustainable agriculture movement. Paper presented at the annual meeting of the Rural Sociological Society, Portland, OR.

Whatmore, S. and Thorne, L. (1997). Nourishing networks: Alternative geographies of food. In D. Goodman and M.J. Watts (Eds.), *Globalizing food: Agrarian questions and global restructuring* (pp. 287-304). London: Routledge.

Wheeler, R. (1996). Special issue on 'Empowerment', *Local Government Policy Making,* 22, 4.

Wilkinson, K. P. (1999). *The community in rural America.* Middleton, WI: Social Ecology Press.

Winne, M., Joseph, A., and Fisher, A. (1997). *Community food security: A guide to concept, design and implementation.* Venice, CA: Community Food Security Coalition.

Wright, W. (1999). *Turning over a new leaf: Socio-economic and political transformations in the burley tobacco commodity system.* Unpublished Ph.D. Dissertation, Department of Sociology, Lexington, KY: University of Kentucky Press.---.
(2003). Fields of cultural contradictions: Lessons from the tobacco patch. *Agriculture and Human Values,* 22(4), 465-77.---.
(2005). Book Review, Civic agriculture: Reconnecting farm, food and community.*Rural Sociology*, 70(2), 276-278.

Wright, W. and Middendorf, G. F (2007). *The fight over food: Producers, consumers, and activists challenge the global food system.* University Park, PA: Pennsylvania State University Press.

Zepeda, L. and Leviten-Read, C. (2004). Consumers' views on local food. *Journal of Food Distribution Research*, 35(3), 1-6.

Linking Small Farms to Rural Communities with Local Food: A Case Study of the Local Food Project in Fairbury, Illinois

S.A. Hultine, L.R. Cooperband, M.P. Curry and S. Gasteyer

The literature on sustainability and local food systems often takes for granted a positive impact on rural communities. This paper attempts to interrogate this notion through a systematic evaluation of a local food system project in Central Illinois. Using case studies and survey research, this paper focuses on an innovative relationship forged between farmers and a local independently owned supermarket in a small rural community, which created an "indoor farmers' market" inside the store. This growing group of farmers committed to providing their community with food grown using sustainable farming practices, while the grocery store owners committed to the project to support local farmers and as a competitive marketing strategy. Although one key measurement of the success of this business relationship is evident in the doubling of local produce sales within the store each year, benefits are accruing beyond the economic impacts. Based on the "Fairbury Local Food Project" we recommend nine key elements of success that can be used to measure impacts and evaluate the success of local food system projects in rural communities.

The predominant forms of agriculture in most of the Midwestern United States are built around large-scale commodity production systems (mainly cash grains such as corn, soybeans and small grains) where most of the inputs purchased for production originate from outside the region of production and the crops are exported outside those regions as well. The resulting economies and social infrastructure of those rural communities provide little to no incentive for creation of a local food system that promotes local production for local consumption. Foods consumed by rural residents travel an average distance of 1,500-2,000 miles (Halweil 2002) and little to none of the food dollars spent within those rural communities stav within the local economies (Meter and Rosales 2001).

In contrast to the current globally based food system found throughout the rural Midwest, locally based food systems encourage production and consumption of diverse foods (fresh fruits, vegetables, meats, poultry, and dairy) within defined geographic regions, often within a 50 – 80 mile radius. The purported benefits are increased availability of more diverse foods; establishment of direct connections between producers and consumers; and revitalization of the infrastructure needed to grow, process, distribute and consume those foods. In a report describing local food initiatives in rural Minnesota, economist Ken Meter noted that these local food ventures have the additional impact of building "connections among community members, and…building wealth that takes root in the local community" (Meter and Rosales 2001, p. 10).

Recent research on rural food systems has highlighted challenges to building local food systems in regions where industrial agriculture prevails. One problem is how to define "local." Several researchers have highlighted that various meanings attributed to "local" food systems can become a source of conflict among scholars and practitioners (Dahlberg, 1994; Hinrichs, 2003). Such attributes encompass travel distance from farm to buying venue, farming practices (environmental sustainability, animal welfare, socially responsible work environment, etc.) and the nature of relationships between farmers and food buyers (farmers' markets, community supported agriculture, wholesale buying of locally produced foods, etc). Geographic region can be as small as a ten-mile radius around a rural community to a multi-state region around a very large metropolitan area. For example, Chicago, IL, considers food produced in Illinois, Wisconsin, Indiana and Michigan as "local," while a small rural town in Central IL considers only the farmers from their town and neighboring towns as "local." Ostrom (2006) found that while meanings attributed to the word "local" differed between farmers and consumers in Washington State, there were areas of overlap within the meanings that provide all stakeholders with a positive image of local food systems.

Additional research in rural communities of Washington State, regarding producer-consumer relationships, identified spatial issues, quality of food, and social relationships as predictors of rural residents' local food purchases (Selfa and Qazi, 2005). Our research furthers these studies by analyzing relationships between not only producers and consumers in a rural community in Central Illinois, but also producer relationships with local government, business owners, and community leaders. In regions where agriculture has traditionally been considered a major component of the rural economy, promoting locally based food systems could be a strategy to reconnect producers to the local community and revitalize the rural economy.

Research on the impact of local food systems on rural communities has presented them as a potential opportunity for re-connecting agriculture and rural communities (Smithers, Joseph et al., 2005). As the scale of agricultural operations has grown and become more industrialized, rural residents have fewer connections to their local farmers. Local food systems can rebuild agricultural infrastructure in these communities, such as new processing facilities or local farmers' markets, that create new opportunities for interaction between rural residents and rural farmers. Smithers (2005, p. 5) points to other research (see Goodman, 2003) that has also focused on local food systems "as social and oppositional movements, without explicit regard for their strategic role in rural development" (Goodman, 2003). While the literature on local food systems provides important theoretical perspectives for building local food systems, the next step is to provide communities with concrete examples of local food system activities. For example, how do communities initiate projects, and what are the opportunities and challenges and operating strategies that provide stakeholders with practical information for starting a project in their own community that emphasizes "their strategic role in rural development?" Our goals for this article are to demonstrate that building local food systems can be an effective strategy for community and economic development in rural communities and to present a model for building a successful local food system in a small rural community.

In this sense we will build on earlier work by Flora (1995) and Lyson (2004) by providing evidence of the role of sustainable agriculture and local food systems in building social capital and improved development potential in rural communities. Both of these authors argued that critical pieces in the development of local food systems are related to the development of social capital— the development of relationships and networks based on values so that the relationship between producers and the community moves beyond market interactions (where the farmers seek to sell at the highest possible price, and buyers seek to buy at the lowest price). Our study seeks to build on the observations of Flora and Lyson through describing an innovative approach to developing a local food system in a small rural community in Central Illinois. We move toward developing a model for communities and practitioners by identifying critical elements of success for local food system projects.

METHODOLOGY

The goals of our overall research project were to document the state of local food systems in central Illinois between 2004-2006 and identify examples of successful local food system activities that could serve as models for other communities. All aspects of this project received approval from University of Illinois at Urbana-Champaign Internal Review Board for Human Subjects. The case study discussed in this paper developed through discussions with several local producers in the region, but also was justified by a trend discovered during our evaluations of several farmers' markets in both rural and urban communities. Surveys at farmers' markets and from a random sample of households (Hultine and Cooperband, unpublished data) identified a general trend of strong support for farmers' market-centered local food systems in the urban areas of Champaign/Urbana, Bloomington/Normal and Peoria, with weaker support for farmers' markets in rural areas. The one outlier of this rural trend was a local food systems initiative in the community of Fairbury, in rural Livingston County, Illinois.

Case study documentation of the Fairbury local food project began in late 2004. We were interested in how the project developed during the three-year period, thus a more qualitative approach was utilized to document participants' reactions and impressions of the project.

One quantitative method was employed to document consumer preferences in the community. Intercept surveys of customers at the local supermarket featuring locally grown foods (Dave's Supermarket) were conducted on September 21 and October 1, 2005. Surveys were conducted during four one-hour segments between 10:00 a.m. to 6:00 p.m., considered peak shopping hours by the owners. Shoppers were intercepted in the produce section of the store, and asked to complete a three-page survey questioning their shopping habits, preferences and attitudes about their community. Seventy-one consumers completed the survey. Results were tabulated and analyzed using SPSS statistical software.

The participating farmers, Dave's Supermarket co-owner, and owners or managers of several other Fairbury businesses that purchased locally grown products from the farmers completed short open-ended surveys in 2005 and 2006 on the impact of the project, including sales figures, production costs, thoughts on opportunities and challenges, and the benefit to the community. The farmers' group held planning and brainstorming meetings several times per year to discuss the project and methods for improving production, marketing and recruitment of new farmers. Participant observation techniques were used to observe and document the interactions among the farmers and the content of meeting discussions and to study volunteer workdays and special events/presentations at local farms and the grocery store since 2004.

Interviews with key participants in the project, including the manager of Dave's Supermarket and the three lead farmers, were conducted on four separate occasions beginning in July 2005. Phone interviews with the Fairbury mayor and a member of the Fairbury Economic Development Committee were conducted in June 2006. These interviews included discussions of background for the project and collaboration, plans and strategies for the future, ideas for maintaining a successful collaboration, attracting more consumers and farmers, and the benefit to the community.

THE FAIRBURY LOCAL FOOD PROJECT

Fairbury, Illinois, a community of nearly 4,000 residents, is located approximately 115 miles southeast of Chicago (see Figure 1). Table 1 displays demographic characteristics of Fairbury. Livingston County has a population of only 40,000, and although only six percent of county employment is in agriculture (U.S. Census Bureau 2000), nearly 90 percent of land in the county is used for agricultural purposes (U.S. Census of Agriculture 2002). In 2004, two farm families in Fairbury initiated a collaboration with a local community business, developing an indoor farmers' market inside the independently owned local grocery store. Dave's Supermarket is the only grocery store within a 15-mile radius of Fairbury and plays a central role in the business district of the community.

Figure 1. The case study community of Fairbury, Illinois, is indicated with the black dot

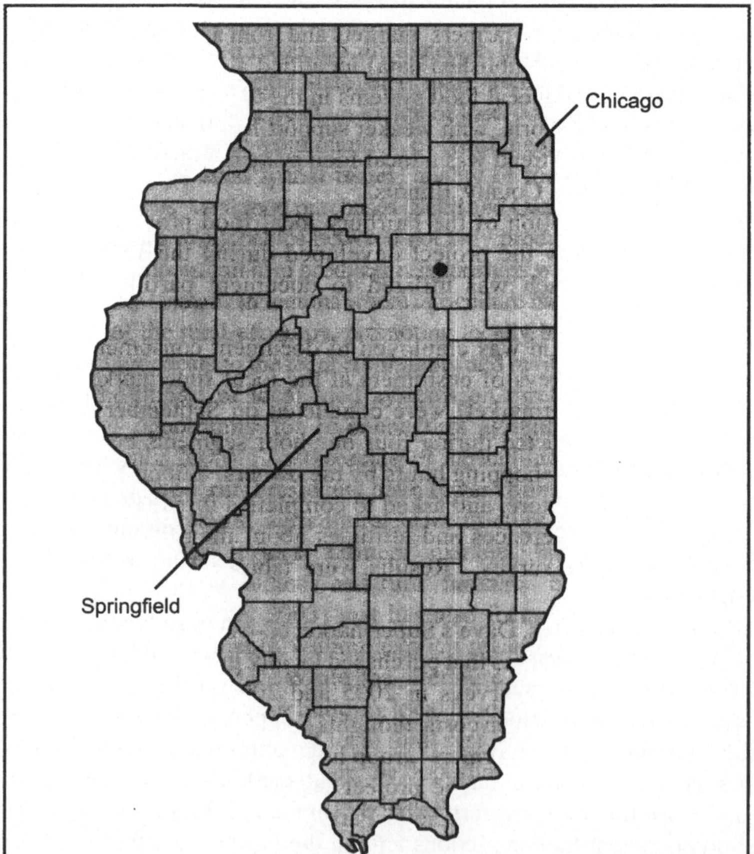

Table 1. Demographic data for Fairbury and the State of Illinois

	Fairbury	Illinois
Population (2000)	3,968	12,419,293
Median household income (1999 dollars)	$41,298	$46,590
Families below poverty level	3.3	7.8
Bachelor's degree or higher	15.4	26.1
Median age	38.9	34.7
Average household size	2.48	2.63
Mean travel time to work (min)	22.1	28

Source: U.S. Census Bureau, 2000 Census

Initially, the co-owners of Dave's Supermarket were reluctant to purchase food products from the local farmers because of uncertainties about the demand for the local products. During a second conversation between the farmers and storeowners, the idea of the indoor farmers' market developed mutually by both groups. Neither group investigated the existence of similar arrangements in other communities. The arrangement that the farmers and store owners developed was to create an "indoor farmers' market" in the produce section of the store. The business arrangement was very informal at the start, with the shared idea that the farmers would provide whatever produce they could for the store, and that they would start small to determine how the arrangement worked for both the store and the farmers. The grocery store provided the farmers with a small amount of shelf space in the store, and the farmers were responsible for stocking the shelves, determining prices, and ensuring a quality presentation of their products. The store provided barcodes that identified each product to the farm where it was produced, and the products were scanned through the checkout lines, the same as any other store products. The store received a 20 percent cut from the sales of the local produce, with the rest of the sale going directly to the farmers. Monthly, the store sent a check to the lead farmer with total sales for each farm. The store also provided advertising for the farmers' products.

As the "local foods project" has grown, the store expands shelf space for the local produce, as long as the products are displayed well, and the farmers remove any products that are not sold in time. Although the farmers are not certified organic, they have all committed to growing their products using organic methods and promote their products as "naturally grown and chemical-free." Each new farmer who joined the project agreed to use organic or natural production methods as well. Initially, local products sold in the store included lettuce, tomatoes, potatoes, and other vegetables, and eggs. As the number of farmers has increased, so has the variety of products available, including a variety of heirloom vegetables, eggs, chicken and beef, and more unusual items such as leeks and paw paws (a native fruit). More traditional products such as salad mixes, tomatoes, and potatoes have higher sales, but as word about the project has spread and positive reviews of the quality of food sold by the farmers have been received throughout the community, the farmers have seen higher sales of the more uncommon and unusual products they provide in the store.

At the project's inception, the two lead farmers sold their products in the store from May through October, along with an additional producer with a small market garden who sold products only for half the season due to crop damage from a hailstorm. Total sales in 2004 were $850. Total sales in 2005 with five participating farmers were $2,009, and more than doubled in 2006 (with 11 farmers) to $4,509. Sixteen farms were expected to participate in the 2007 season.

Although the farmers and the Dave's Supermarket owners are the key stakeholders in this project, their business relationship has benefited from their other relationships throughout the Fairbury community – such as the involvement of the lead farmers and Dave's owners in the chamber of commerce and the lead farmers' work with their non-profit farm foundation board. The business relationship between the farmers and Dave's Supermarket has also served as a catalyst for new relationships within the community. Figure 2 illustrates the key players and the nature of their relationships within the local food systems project in Fairbury. For example, other retail businesses have requested local produce from the farmers after seeing the success of the local food projects at Dave's, while consumers are interacting with both the farmers and the supermarket by providing their input on which products they would like to buy locally from the supermarket. In the following sections, we will describe in more detail four of these constituencies and their relationships within the community: the participating farm businesses, Dave's Supermarket, consumers, and local Fairbury officials.

Figure 2. Local Food System Relationships in Fairbury

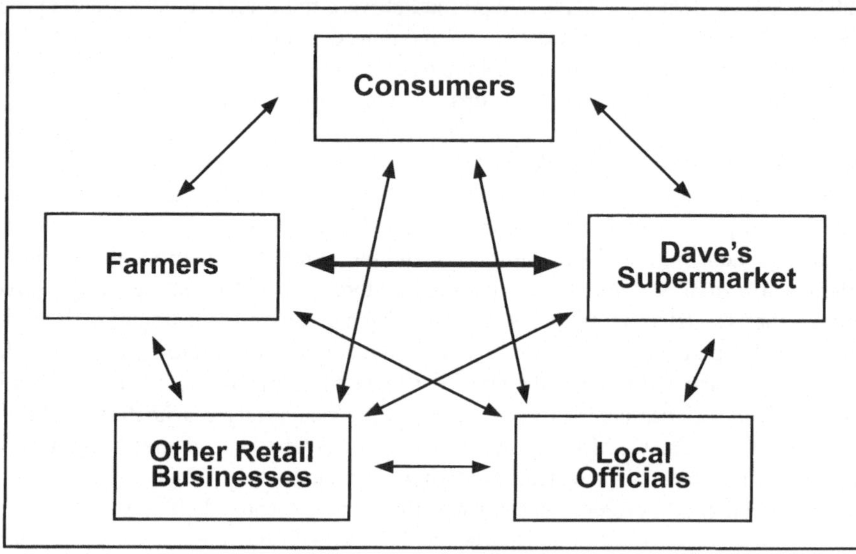

The business relationship between the farmers and Dave's Supermarket is the most strongly developed, but new relationships are developing between the other stakeholders in Fairbury as a result of the 'Fairbury local food project.'

Farm Businesses

Two entrepreneurial farms initiated the relationship with Dave's Supermarket. One of the farm operators included a recent college graduate who returned to his family's cash grain (corn and soybean) farm. He was allocated two acres to raise and sell chickens, eggs, and vegetables to local markets. The other farm is the oldest farmstead in Livingston County and is managed by a descendant of the original family and his wife (this couple will be referred to as the lead farmers for the remainder of this article). The mission of the historic farm is to provide agricultural education and a demonstration of the viability of small-scale sustainable agriculture for the region. Both farmers were relatively new to small-scale, intensively managed farming and direct-market agriculture at the inception of the project, although each had backgrounds and education in gardening and horticulture. These farmers were idealistic in their goals for their farms, which included using organic production methods and providing fresh, healthy, locally grown produce for their communities. Both of these farm families had off-farm jobs, but were raising their direct-market food products on two acres or more. The farmers were interested in finding a new market that was closer to home, rather than driving more than one hour to sell at large urban farmers' markets.

The lead farmers persistently recruited and encouraged new participants by inviting prospective farmers to visit their farm, attend workshops on direct-marketing, and sharing information about their own experiences with direct sales. The farmers also supported involving youth in the project, and welcomed several high school children who became interested in growing and selling produce in the store. These teenagers participated with their families, raising a variety of vegetables during the second year of the project. The producer group was very supportive of these young farmers and worked to encourage several other school-age children who started selling turkeys, eggs, and vegetables in the store in the third year. The farmers saw this as an opportunity to help young people become interested in sustainable agriculture and local food systems.

The producer group has grown each year: three farmers in 2004, five in 2005, eleven in 2006, and sixteen farmers participated in the 2007 season. The group is now predominantly small market gardeners, with only four participants who are full-time farmers. Because of this predominance of market gardeners who have limited production and marketing experience, the farmer group has decided to grow the project slowly and promote it as not only a moneymaking outlet, but also an educational opportunity for the producers. This allows the participants to experiment with produce varieties and quantities, production methods, and learn as they go about production and marketing strategies.

The lead farmers regularly convene planning meetings so that all of the participating farmers can meet, discuss ideas and concerns, and coordinate products and quantities. Besides providing an educational component at these meetings, they are also strengthening the relationships among participating producers by discussing goals and challenges, sharing information and leadership roles, and providing encouragement. At these meetings, producers reinforce their commitment to high quality standards for the products they sell in the store as well as their commitment to using organic production methods. By allowing all participants to have a role in deciding production and marketing decisions, these meetings solidify the values of the group and strengthen the level of participation and commitment from each of the producers.

At a recent planning meeting, participants shared their goals for their participation in the Fairbury local foods project. Nine of the farmers discussed interest in the project as a hobby or enjoyable activity, as an opportunity to work with family, and a source of extra income. Four of the farmers discussed their hope that they would be able to grow their farming businesses to generate more income from farming, so that they could eventually leave or change their current jobs to farm full-time. One producer highlighted the long-term nature of these goals: "We both have full-time jobs, so it's difficult to have time to grow our [farming] business bigger than it is now. We would like for this to be a full-time venture for both of us, but we have to make sure it's profitable first."

Since the project's inception, the farmers have promoted to their community the shared goals of providing food for local consumers and a new market for small family farms. The relationship with Dave's Supermarket has strengthened through improved communication and higher levels of trust; the continued positive support from Dave's is an important aspect of the business relationship. Recently, one of the farmers discussed the idea of the farmers' group becoming a major presence within the store in the next five years, diversifying the producers and products available in the store, and continuing to receive positive comments from the community. With one exception, each of the participating producers has income outside of farming, so that they are not dependent on profits from the Dave's Supermarket sales as their only income. This allows the producers to focus on standards of quality, develop a base of consumers, and experiment with products and advertising methods. Only two farmers have withdrawn from participating in the Dave's project. One of the two farmers who initiated the dialogue with Dave's Supermarket has farming as his sole means of income, and found it necessary to search out more profitable sales outlets for his direct-market food products. The other participant withdrew because of increased time constraints at her place of work that limited her ability to commit time to her garden and the project.

In late 2006, the farmers began discussing the creation of a cooperative to continue formalizing their relationship with Dave's Supermarket, and also to begin addressing some of the new potential markets that were beginning to arise from their success at the supermarket. Several other grocery stores and restaurants in Livingston County have expressed interest in buying local products from the farmers, and there are new opportunities to sell to restaurants in Chicago.

As the farmer group is still small and production capacity is limited, the group felt that becoming a more organized entity would help them to raise capital funds and develop a market presence to address these new opportunities. After several meetings with a University of Illinois Extension educator and local lawyer, the group decided that forming a limited liability company was a better option than forming cooperative, because of the flexibility and limited operating requirements. Each participating farm pays a membership fee, and then all sales made using the LLC label will be covered by a group insurance policy. Participating farmers will also benefit from the marketing presence that the group will generate, from new food markets that require greater quantities than an individual farm can produce, and the opportunity to share equipment and other business costs among the group members. Core values of the group include: production using chemical free methods as outlined in the USDA National Organic Program organic standards, no reselling of products, providing educational and mentoring opportunities for producers and consumers to share knowledge of agriculture in their community, and promoting a collaborative environment for the producers to share and learn from each other. The group incorporated the LLC for the 2007 season.

Retail Outlet

Dave's Supermarket opened in 1950 as an independent grocery store and remains under the ownership of the founding family. The store is one of the largest employers in Fairbury, with 145 employees. The next closest full-service grocery store is located in Pontiac, approximately 17 miles from Fairbury. In addition to selling groceries, the store's services include a deli serving lunch and dinner, catering services, a website and quarterly newsletter, dry cleaning, film developing, floral center, and bakery. The owners are committed to serving the community, as members of the chamber of commerce and by making contributions to local organizations and causes.

The technical aspects of the business arrangement between the farmers and storeowners allow the supermarket to support the local farmers with minimal investment and provide an important public relations benefit to the store. Dave's Supermarket recently faced competitive pressure from a new Wal-Mart in a neighboring community and attempted to develop new strategies to help differentiate their store to better compete with big-box stores. Dave's promotes "We Buy Local" and their relationship with the local farmers as a unique opportunity that is not available at other stores.

A benefit of the local ownership of Dave's Supermarket is that the storeowners are able to make quick decisions about ideas and new projects. The owners are able to meet with producers to discuss ideas and concerns. They are also willing to assume some risk for the project, and experiment with the producers to find the most successful process for implementation of the local food project.

One of the biggest challenges that the producers and store owners faced in the initial years of the project was maintaining open communication to ensure mutual expectations were met. During one interview with the store co-owner during the second season of the project, he mentioned his overtures to the farmers to advertise their products weekly, but had not received any responses from them. He was also unsure why the farmers hadn't been interested in participating in a tent sale during that season. Once the farmers shared their growing challenges that season, due to weather and pest problems that limited their harvest quantities, the store owners were very understanding and supportive. As the farmers' group became better organized, the level of communication between the farmers and the store owners became more formalized and systematic as well, limiting the miscommunications and misunderstandings between the two groups.

One of the biggest contributions that Dave's Supermarket provided to the local food project is advertising and publicity. The store advertises the local products in their weekly newspaper advertisements, highlights the products at special sales events, hosts outdoor tent sales for consumers to meet the farmers, and discusses the project with consumers both inside the store and throughout the community. Inside the store are several signs, including the "Buy Fresh, Buy Local" materials from the non-profit organization *Food Routes* (www.foodroutes.org), and a sign "Dave's supports local farmers," along with information about each of the participating farmers. Each year, the co-owners have encouraged the farmers to use more advertising methods and suggested trying more tasting demonstrations for 2007. Several additional events have helped to publicize the Fairbury local food project to the community, including a stop on one of the participating farms by the FarmAid tour in 2005. Several articles and press releases in local, regional, and national newspapers, including articles in the Chicago Tribune and New York Times, have also sparked interest in the project.

The store owners have consistently been optimistic about the potential for the project and believe that sales will continue to grow in coming years. The store's participation in the project is not a significant risk for their store, although their main risk is in the lost profit opportunity from sales of non-local food that could be stocked on the local producers' shelves. However, the co-owner notes that the benefits from the public relations opportunity gained by working with the local farmers are worth the lost income from produce sales. The store owners know that their consumers are supportive of local products and appreciative of the uniqueness of the local food sold within the store. With the development of the limited liability company to support the producer group, Dave's has agreed to buy local products exclusively from the LLC, and will refer any farmers who are interested in selling to the store to meet with the LLC. This willingness to give the producer LLC an exclusive opportunity for store sales demonstrates the transition to a more formal business relationship between the store and the producers.

Consumers

The intercept survey of customers in Dave's Supermarket, as described in the methods section, provided a profile of shoppers at the store. Seventy-one customers were surveyed to understand their shopping preferences, habits, propensity to buy the local farmers' products, and their attitudes about their community. Table 2 highlights key findings from the survey.

Ninety-five percent of shoppers surveyed in Dave's Supermarket did their primary shopping there. Forty percent of those consumers surveyed resided in Fairbury, while all other respondents lived within 35 miles of Fairbury (with one exception- a consumer visiting Fairbury from Urbana). Most were the primary shoppers for food in their household (91 percent), 93 percent were women, and the average household size of the customers was 2.8 people. Nearly half of the customers earned more than $50,000 in household income per year, and 38 percent had an associate's degree or higher. Customers were very supportive of Dave's Supermarket, as one customer commented: "I appreciate Dave's. It's definitely a vital part of Fairbury. It's wonderful to have this quality of store close to home!"

Table 2. Highlight responses from the Dave's Supermarket consumer intercept survey. These results were tabulated from two intercept surveys conducted in the supermarket on Sept. 21 and Oct. 1, 2005.

	Dave's Supermarket Survey Respondents
Number of shoppers surveyed	71
Surveyed shoppers who do their primary food shopping at Dave's Supermarket	95%
Fairbury residents	40%
Primary shopper for food in household	91%
Female	93%
Average household size	2.8
Annual household incomes greater than $50,000	48%
Associate's degree or higher	38%
Average monthly expenditures on groceries	$345
Average # of shopping trips to Dave's Supermarket per month	6
Have you purchased products from the local farmers in the store?	65% yes
Would you recommend the farmers' products to your friends?	98% yes
The local products available in the store are fresh and high quality.	90% agreed
I purchase products from the local farmers' shelves to support local farms.	87% agreed
I would buy more products from the local farmers if available in the store.	85% agreed
The local products are priced fairly.	80% agreed
The business relationship between the farmers and Dave's Supermarket is good for the community.	85% agreed
Which advertising method is best for influencing your purchases of locally grown food products?	Prior purchase of a local product: 80%; Word of mouth: 65%

A majority of customers surveyed had purchased the local products in Dave's previously (65 percent). Nearly 100 percent of customers who purchased products from the local farmers in the store said they would recommend the farmers' products to their friends, and 90 percent agreed the farmers' products are fresh and high quality. The consumers agreed that they purchase products from the local farmers' shelves in the store to support local farmers (87 percent) and that they would buy more local products if available (85 percent). Eighty percent agreed the local products were priced fairly, and 85 percent believed that the business relationship between Dave's Supermarket and the local farmers was good for the Fairbury community. Several of the farmers received letters and phone calls from consumers, complimenting them on the quality and thanking the producers for providing local consumers with produce grown without chemicals. One letter in particular, from a cancer survivor, credited the local food she purchased from the farmers as helping her stay healthy. Word of mouth is one of the best advertising methods for the project, with 65 percent of customers surveyed in Dave's Supermarket responding that word of mouth i s one of the most influential methods for learning about the local food products. A prior purchase of local food products in the store is also effective in influencing customers to continue purchasing the local products (80 percent of respondents).

Customers suggested products that they would be willing to purchase from local farmers inside Dave's, including eggs, meats, and a variety of additional fruits and vegetables that had not been sold by the farmers. This information from the consumers encouraged a new young student to begin selling eggs inside Dave's, and several new producers began working with the store owners to sell meat inside the store.

Local Leaders

Fairbury has 200 businesses within its boundaries and a small but active downtown. The community is currently using several economic development strategies such as creating an economic development committee (EDC), improving the Chamber of Commerce website, and focusing on accommodations, industry and commerce. The EDC is also considering expanding city boundaries to draw several outlying businesses in to the city to generate more sales tax.

Interviews with the community's mayor and an economic developer revealed a high level of awareness of and support for the Fairbury local food project. The chairperson of the EDC discussed the impact of the project with the EDC and recognized the value of the farms as a tourism opportunity for the community, as visitors to the local farms must drive through Fairbury. She appreciated the business relationship between the farmers and the store and the opportunity for local consumers to see products with a Fairbury label. A growing population within the Fairbury community is interested in organic produce and this project is serving that growing market. The publicity that the farmers and the project have received also benefits the community by drawing positive attention to Fairbury.

Although the mayor has had limited contact with the farmers involved in the project, he has purchased many of the products, and has heard that customers are pleased to have the local products available in the store. The City is committed to helping agri-business ventures and could do more to help promote the project to customers in nearby communities as well. The mayor is hopeful that the project will continue. He frequently mentions the project to businesses who are considering locating in Fairbury, and considers the project good public relations for the community. He believes that the mindset of the Fairbury community is important in making the project successful – Fairbury is unique in the level of pride and support for the community and its downtown businesses. We compared survey responses from Dave's customers who live in Fairbury and residents in other rural communities (from unpublished data collected from farmers' market surveys and a random sample household survey) on the question "how would you rate your community as a place to live, work, visit and do business?" The results supported the mayor's claim, as Fairbury residents had the largest proportion of excellent ratings for each indicator, shown in Figure 3.

As the project is still new to the community, the local leaders interviewed have as of now provided their support through their own purchases at the store, and through verbal support and recognition to their organizations and colleagues. But both are considering more extensive opportunities to contribute to the local food project and to small-scale agriculture in their community. Some of the ideas they are currently developing include holding seminars for producers and local entrepreneurs to answer questions and share ideas that would create an opportunity to build more small scale farms and businesses that would support the local food system within the community. Local government officials are also interested in providing educational materials to educate consumers on the variety, uses, and prices of unique local products available from the farmers. The chamber of commerce is working to expand its boundaries to include rural farmers, so that the farmers would be eligible for the chamber's small-business loans and grants. Both the mayor and EDC chairperson believe the Fairbury local food project could be replicated in other communities. Their additional strategies for providing local government support to local food system projects are included in Table 3.

Figure 3. Consumers rated their communities as a place to live, place to conduct business, place to work and place to visit

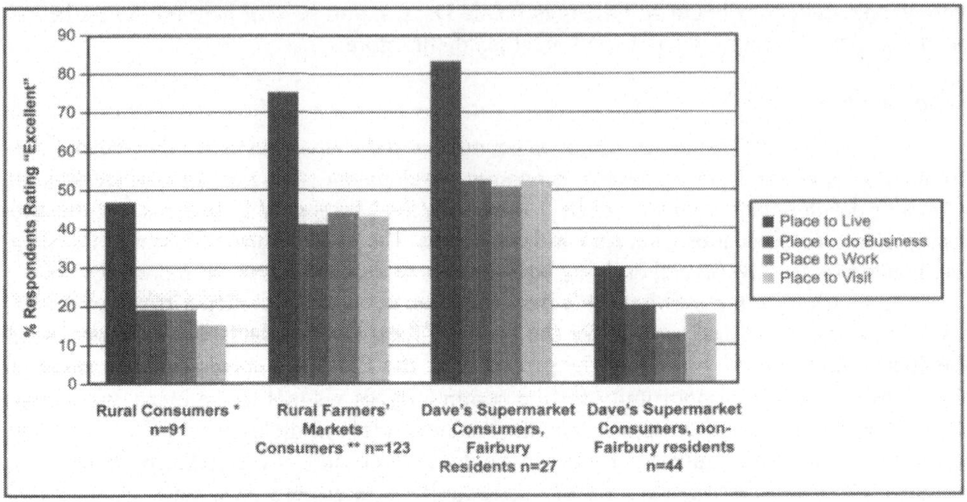

* Unpublished random sample survey of households in 22 central Illinois counties.

** Unpublished intercept surveys at six central Illinois farmers' markets.

Table 3. Recommendations from Fairbury Elected Officials for Building Local Food System Relationships.

* Develop educational materials targeted to rural residents that promote the benefits of locally grown produce.
* Hold workshops that promote or help establish entrepreneurial agricultural businesses.
* Encourage collaboration among growers to provide a variety of products.
* Include rural businesses (including farms) within the chamber of commerce.
* Encourage appreciation among residents of alternative agricultural operations within the community.
* Support relationships between producers and food buyers as "business to business relationships" through the local chamber of commerce.

Key Elements for Success

Based on our case study of the Fairbury local food project, we identified key elements that we perceive influenced the success of the local food project. We present the following nine key elements for successful community-based local food systems as a potential model against which other local food system ventures could be tested.

Catalyst Farmers

The farmers who initiated the local food project in Dave's Supermarket are motivated, persistent entrepreneurs willing to take risks and grow the project slowly over time. These farmers are willing to step beyond their own farming operations to play an active role in the development and maintenance of the relationship with Dave's Supermarket. It will be important for emerging local food markets in rural communities to recruit farmers who are energetic, enthusiastic, and willing to contribute their creativity and expertise to the development of local food markets. They were also willing to mentor young and beginning farmers. Markets that incorporate the input and experience of entrepreneurial farmers will be more likely to serve the needs of the participating farmers, building a stable base of producers and products that will in turn provide a more consistent shopping experience for consumers.

Independent Retail Outlet

An independently owned retail outlet, such as Dave's Supermarket, where owners are community members and decisions are made locally, can provide a better opportunity for farmers and business owners to build a strong business relationship. In a discussion of food retailing in the U.S., Guptil and Wilkins (2002, p. 45) suggest that food retail businesses develop distinctive "identities with relation to the communities they serve" and that local producers should work with these businesses to tailor their product offerings to represent these locale-influenced markets. Local decision-making can allow food business owners (groceries, restaurants or specialty stores) to experiment with new opportunities for buying and selling locally grown food, and allow for better communication among the business owners and producers to address challenges and build on successes with a quicker response time.

Communication

Initial communication between the Fairbury farmers and Dave's Supermarket was informal, which posed several challenges for the project stakeholders in the first and second project years. As the relationship between the farmers and Dave's owners developed, they have created a more formal communication network, which has helped to limit misunderstandings by addressing issues more quickly between the stakeholders.

Communication must extend beyond the producer and business owner or food buyer. There must be a strong message conveyed to consumers and the community about the goals of the project, and open communication networks with local government and other community leaders can strengthen the level of support for the local food project within the community. While more formal interactions between producers and business owners can enhance the business relationship, using a variety of techniques for communicating the marketing message can be important as well. Both the farmers and the food business owners have used their relationship with community leaders to promote the project as well. The community leaders have become informal spokespeople for the project. While using more formal methods such as radio and newspaper announcements or chamber of commerce presentations can be an effective strategy, the value of informal communication – word of mouth advertising, for example - should not be overlooked.

Long Time Horizon to Achieve Success

Feenstra suggests that for local food system projects to maintain their spaces in a community, stakeholders must "allow time [for projects] to grow roots (2002, p. 102)."

In the Fairbury local food project, the participants can forgo short-term profits from the Dave's Supermarket project because they have other sources of income. They have made an effort to not only develop a new market for their local food, but also to advance the goals of building community relationships and drawing customers to other downtown businesses. If stakeholders can establish long-term goals or visions for local food projects (Meter 2003) and appreciate small successes in the short-term, the project participants have a greater opportunity to experiment and build on lessons learned as the project grows.

Leadership

Economist Ken Meter recommends democratic leadership as a critical element to build community-based food systems (Meter 2003). Throughout the evolution of the Fairbury local food project, participants have been willing to take responsibility for leading the group in decision-making, coordinating accounts and convening planning meetings. The producers have also been willing to share responsibility – in 2007, the accounting task will be handled by a new member of the producer group (the lead farmers had assumed this role for the first few years of the project). Several producers have created committees to research and present ideas on the formation of the limited liability company, and to meet with representatives from potential new markets. Collaborative, positive leaders can provide local food projects with a strong sense of direction and stability. An enthusiastic and knowledgeable manager or leader who is committed to the project will draw support and develop positive relationships within the community. Campbell (1997) advocates that collaborative leadership and the training of new leaders was a key aspect for the success of the California Alliance for Sustainable Agriculture.

Community Cohesiveness and Pride

The Fairbury mayor commented that the high level of community pride among Fairbury residents was one reason for the support of the downtown business district and the local food project. Communities that place an emphasis on strengthening relationships among community members and promoting the positive aspects of their community may be more willing to support new markets for locally grown food. Vibrant downtown business districts demonstrate that local residents are willing to support local businesses and are likely to support new ventures within their community. Community members who appreciate their community can also provide additional advertising for local food projects by sharing their recommendations to other community members.

Supportive Local Officials

Both local leaders interviewed from Fairbury gave highly enthusiastic and supportive comments about the Fairbury local food project. They identify the project as an opportunity to enhance Fairbury's image, draw in tourists, and promote the supportive business environment in the community. These leaders can see the potential in building local food systems as a strategy for community and economic development and have been willing to work with the local food group. There are many opportunities for local government leaders to support local food system projects, ranging from promoting the project within the local community, to providing business incentives such as low-interest loans, grants, or other small business funds. Planners can also contribute skills for building a community's food infrastructure to include local food systems (Campbell 2004). Engaging local government leaders as stakeholders in local food system projects can add an additional level of validity to a new project.

Invested Consumers

The farmers and store owners in Fairbury made an effort to address the concerns and interests of consumers in their local community. Consumers with preferences for organic or locally grown produce, who speak highly of their community, are the best proponents for locally grown food, and their continued support through purchases and recommendations can build the customer base for local producers. In suggesting a plan for building local food systems throughout the state of Kentucky, researchers promoted "reaching out to consumers as partners" as a strategy for success (Plath and Blobaum 2003, p. 34). Understanding the shopping behaviors and preferences of local consumers can help local food system projects to promote aspects of their businesses and products that address consumers' demands. Consumers who are invested in their communities and willing to support local businesses are vital to the success of local food system projects.

Location

In Fairbury, Dave's Supermarket has the benefit of a centralized downtown location and an established reputation as a high quality grocery store. Finding an attractive and competitive location for local food system projects, from farmers' markets to retail businesses, is key to attracting customers. Central downtown locations can be successful if they serve as a draw for consumers. Making connections between the local food market and other local business or organizations, such as hosting promotional events or sharing advertising, is another means to enhance the location aspects of a market.

CONCLUSION

Smithers argues that "there may be innovations in agriculture outside the dominant industrial model that would serve to promote engagement and mutual sustainability (2005, p. 13)" and suggests that food purchasing initiatives are one opportunity for rebuilding agriculture within rural communities. The intention of this paper is to build on the theoretical frameworks in previous research by providing a concrete example of a local food system initiative in a rural community that played a positive role in building new social networks and creating a new image of agriculture for the community. As demonstrated through this case study of the Fairbury local food project, local food system projects can provide rural communities with unique opportunities for improving markets for local farmers, developing new business relationships in the community, and building community connections. Each rural community will require an individualized approach to finding the best opportunity to build local food systems in their community, but it is clear, as asserted by Flora (1995) and Lyson (2004), that keys to success are related not just to product characteristics and venture capital, but to the existence and development of social capital, as well as cultural capital in the form of community attachment that favors local food systems. We have spelled out above the elements that were important in this alternative case of local food system development in rural Fairbury, Illinois. Further research is needed to test these elements as a framework for collecting data on local food system projects and as a tool for evaluating the impact of local food systems projects as community and economic development.

REFERENCES

Campbell, D. (1997). Community-controlled economic development as a strategic vision for the sustainable agriculture movement. *American Journal of Alternative Agriculture* 12 (1): 37-44.

Campbell, M. C. (2004). "Building a Common Table: The Role for Planning in Community Food Systems." *Journal of Planning Education and Research* 23(4): 341-355.

Dahlberg, K. A. (1994). "Localizing Food Systems." *The Neighborhood Works*(Feb/March): 14.

Feenstra, G. (2002). "Creating space for sustainable food systems: Lessons from the field." *Agriculture and Human Values* 19: 99-106.

Flora, C. B. (1995). "Social capital and sustainability: Agriculture and communities in the great plains and corn belt." *Rural Sociology and Development: A Research Annual* 6: 227-246.

Goodman, D. (2003). "The quality `turn' and alternative food practices: reflections and agenda." *Journal of Rural Studies* 19(1): 1-7.

Guptil, A. and J. L. Wilkins (2002). "Buying into the food system: Trends in food retailing in the US and implications for local foods." *Agriculture and Human Values* 19: 39-51.

Halweil, B. (2002). Homegrown: The Case for Local Food in a Global Market. *Worldwatch Papers*. T. Prugh. Danvers, MA, Worldwatch Institute: 86.

Hinrichs, C. C. (2003). "The practice and politics of food system localization." *Journal of Rural Studies* 19(1): 33-45.

Lyson, T. A. (2004). *Civic Agriculture: Reconnecting Farm, Food, and Community*. Medford, MA, Tufts University Press.

Meter, K. (2003). Food with the Farmer's Face on It: Emerging Community-Based Food Systems.

W. K. K. Foundation: 62.
Meter, K. and J. Rosales (2001). Finding Food in Farm Country, Crossroads Resource Center: 68.
Ostrom, M. (2006). "Everyday Meanings of 'Local Food': Views from Home and Field." *Community Development: Journal of the Community Development Society* 37(1): 65-78.
Plath, P. and R. Blobaum. (2003). "Bringing Kentucky's Food and Farm Economy Home." from www.foodroutes.org/doclib/238/Bringing+KY+Home.pdf.
Selfa, T. and J. Qazi (2005). "Place, taste, or face-to-face? Understanding producer-consumer networks in 'local' food systems in Washington State." *Agriculture and Human Values* 22: 451-464.
Smithers, J., A. E. Joseph, et al. (2005). "Across the divide (?): Reconciling farm and town views of agriculture-community linkages." *Journal of Rural Studies* 21 (3): 281-295.
U.S. Census of Agriculture. (2002) Accessed at www.nass.usda.gov.
U.S. Census Bureau. (2000) "American Factfinder." Accessed at www.census.gov.

Linking Sustainable Agriculture and Community Development:
The Lowcountry Food Bank's Use of Locally Grown Foods

Kenneth L. Robinson, Kathleen K. Robinson, Carlos Carpio, and David Hughes

---

Food banks throughout the South have historically helped meet the nutritional needs of low income families, but are not often involved in community development efforts aimed at creating sustainable agricultural situations. In this paper we seek to critically evaluate a food bank's efforts to enhance small- scale farm operations, increase the quality and quantity of their products, and enhance distribution and use of their product to the emergency food assistance system in Lowcountry South Carolina. The discussion is organized by reviewing the results of the project according to the three primary goals of sustainable agricultural programs.

A growing movement emerged during the 1980s that questioned the role of the agricultural establishment in promoting practices that contribute to social and economic conditions in rural communities. Over the years this movement for sustainable agriculture has gained widespread attention and increasing support. Advocates suggest that sustainable agricultural activities promise innovative and economically viable opportunities for growers, laborers, consumers, policymakers and many others in the entire food system (University of California, 2007; Lyson, 2004). Specifically, sustainable agricultural activities strive to integrate three different goals. First, they strive to use natural resources in a way to promote people's ability to achieve or sustain health. Second, they seek economic profitability while seeking a healthy environment through careful, innovative management practices. Third, they strive to address or correct, as well as do no further harm to, the social conditions, and they seek equitable access by all to market, environmental and community resources. Food banks throughout the South have historically helped meet the nutritional needs of low income families (Duffy et al., 2002; Molnar et al., 2001; Smith and Hoerr, 1992), but are not often involved in community development efforts aimed at creating sustainable agricultural situations. Indeed, it would take alterations in traditional food bank operations in order for a food bank to play a major community development leadership role in sustainable agricultural activities in a region. Thus, few examples can be found in the United States which link food bank activity to local sustainable agriculture despite the potential for providing a healthier food supply to low income residents as well as creating improved small-scale farm operations and revenues.

An example of one effort in coastal South Carolina includes the Lowcountry Food Bank (LCFB). The LCFB is a food distributor that collects, maintains and distributes more than 10 million pounds of food annually to a grassroots network of 247 member agencies providing hunger relief services (Lowcountry Food Bank, 2006). In 2005, the LCFB started a Growing Foods Locally (GFL) program to support local farmers, create new food distribution and marketing channels, and improve the quality of the food that is distributed to the agencies' clientele.

In this paper we seek to critically evaluate the first year's efforts of GFL to enhance a few small- scale farm operations, increase the quality and quantity of their products, enhance distribution and use of their product to the emergency food assistance system in Lowcountry South Carolina. We highlight the nutritional benefits to low income residents who are reliant on the emergency food assistance system for some or all of their daily food intake. The changed role of the local food bank is also examined. The discussion is organized by reviewing the results of the project according to the three primary goals of sustainable agricultural programs.

Literature Review

Through the GFL project, the LCFB and Clemson University aim to significantly contribute to a more sustainable environment for participating small scale agricultural enterprises and simultaneously transform the emergency food assistance system in the Lowcountry of South Carolina into a system that becomes an economic engine for selected small scale enterprises and provides more nutritious food, some of which comes from participating farms. This approach is consistent with a small but growing body of theory and research that has focused attention on small firms, regional trade associations, industrial districts, and local entrepreneurs as potentially important, though often neglected, agents of development (Piore and Sabel, 1984; Porter, 1998; Porter, 2004).

The examination of the organizational embeddedness of small-scale, locally-controlled, economic enterprises and its association with community well-being has historically focused on the production sector, namely agriculture (Goldschmidt, 1946) and manufacturing (Mills and Ulmer, 1946; Piore and Sabel, 1984). The underlying premise is that small-scale production is tied to place by social and economic relationships. It suggests that a dense network of local institutions and organizations, including churches, retail enterprises, and voluntary associations, among others, serves as a glue that binds people to place.

This research stems from the early work of Goldschmidt (1946) which suggests there may be positive benefits to communities that embrace small-scale, locally-based models of development. In a 1944 study of two farming communities in the Central Valley of California, Goldschmidt found that the quality of economic and social well-being was greater in the community characterized by small, locally owned firms than in the community that was characterized by large, absentee-owned firms. He writes:

> the scale of operations...inevitably had one clear and direct effect upon the community: It skewed the occupation structure so that the majority of the population could only subsist by working as wage labor for others. [This occupational structure]...has had a series of direct effects upon the social

> conditions in the community. ...It means that a large portion of the population has little vested interest – economic or social – in the community itself. (Goldschmidt, 1946, pp. 415-416)

These findings support Goldschmidt's hypothesis – large-scale, corporate farming diminishes the quality of life in rural communities – and led him to identify social inequality as the central causal mechanism. Moreover, he downplays the economic contributions of small business development and argues that it is the social networks among government, schools and other social organizations that benefit most from the diffuse resources of a decentralized local economy based on small business.

Research on the emergence of spatially-clustered, local production districts, comprised of loosely-coupled, small firm networks built around a craft form of production (Piore and Sabel, 1984) has helped spawn new interest in local food systems, which also represent the link between small-scale production and community well-being.

Local food systems – sales made by local farmers to local markets – can take a variety of forms including community supported agriculture or subscription agriculture, farmers' markets, roadside stands, u-pick operations, community gardens, and direct sales by farmers to local restaurants, grocery stores and institutions (such as schools and food banks).[2] We examine here the relationship between local farmers and an institution (a food bank). Advocates often point to the Goldschmidt Hypothesis and other more recent analysis, such as the work of Ikerd (2002), in touting the potential benefits of local food systems. The craft production networks of Pierre and Sabel provide another explanation for local food systems.

As discussed in Barkley and Wilson (1995) and consistent with the analysis of Piore and Sabel and others, the Porter model (1985) provides a framework for three strategies that can be used to create and maintain a dynamic spatial cluster. First, firms in the cluster can be low-cost producers by using new technologies that reduce per unit production cost. Second, firms can use a market focus strategy by devoting their efforts in serving a market segment based on location, income category, age, race, or other factors that can lead to differentiate buying habits. Product differentiation is a closely related third strategy whereby customers differentiate the firms' products from those of others.

Advocates of local foods systems explicitly or implicitly follow the Porter-based strategy of market segment and product differentiation. Market segment holds because consumers of farmers' markets, community supported agriculture, and probably other types of local food systems marketing often have higher than average income and education levels, and are viewed as caring more about environmental degradation than the population at large. Place of location (locally versus "elsewhere") also provides a means for product differentiation. Many farmers also like the idea that they are supporting local agriculture, especially small farms, and are helping to retain dollars in the local economy (Brown, 2003; FPC, 2001).

Methodology

A case study method is used to highlight the project, antecedents, context, process and results of GFL in Lowcountry South Carolina (Sterns et al., 1998). The findings are organized around a set of goals of sustainable agriculture introduced by the Sustainable Agriculture Research and Education Program at the University of California-Davis (UC Sustainable Agricultural Research Program, 2007). The information presented is based on field notes, interviews and discussions with farmers, food bank staff and nonprofits affiliated with the food bank. Records kept by farmers, the affiliated nonprofits, and the food bank are also used.

We evaluate the effect of the food bank's use of local foods in terms of enhanced small-scale farm income and sustainability using operations records kept by the food bank. Firsthand information was also obtained by interviewing producers concerning projected changes in income derived from participating in the project.

To evaluate the effect of the LCFB's use of local foods in the food quality and nutritional content of the food offered to the agencies' clientele we measure the changes in the amount of fresh produce that participating agencies received using the project's tracking system created during the first year and the "monthly meal reports" of participating agencies.

Next, we use market information to draw inferences regarding economic impact and the impact on transportation costs. To quantify the effects of the GFL program on the local economy, we used the IMPLAN input-output model building procedure (Alward et al., 1989) to construct an economic model of Beaufort County based on 2004 data. Potential reductions in energy consumption and environmental pollution from meeting local food needs with local products are explored by looking at the point of origin for shipments of produce consumed in the state. Finally, we identify the barriers that have been encountered in implementing the project.

The Growing Foods Locally (GFL) Project

Project Description

The LCFB was a different operation prior to the start of the GFL project. In brief, a food bank purchases items typically found in grocery stores, including food items, and receives donations from a variety of civic organizations or retailers such as churches, individuals, civic associations, and grocery chains, in addition to USDA subsidies. Weekly, bi-weekly or monthly participating agencies will come to the food bank warehouse site(s) to pick up their allotted amount of product. The degree to which a food bank is consciously trying to create healthy inventories is reliant on the motivations of its staff. Fresh produce is not typically available all the time in food banks unless it makes an effort to continuously buy them through a variety of wholesale sources. While some farmers nearby to a given food bank may donate produce, typically these donations are unsolicited, sporadic and done only when a particular farm has excess crop.

The GFL project is an initiative of the LCFB, selected faculty and extension agents from Clemson University, several small scale farms, and 18 community- and faith-based nonprofit agencies. The project simultaneously addresses several community and economic development issues in the Lowcountry of South Carolina.[1] Its aims are: 1) to develop and sustain profitable small scale farming enterprises in Beaufort County, South Carolina; 2) to increase the self-reliance of ten farming enterprises and 25 nonprofit operations working with the LCFB as a food delivery system; and 3) to address the food and nutrition needs of 3,000 low-income consumers in rural Beaufort County and surrounding areas. In essence this initiative seeks a paradigm shift in the way food assistance is done locally.

The GFL project was piloted in Beaufort Country, South Carolina, starting October, 2005. Beaufort County is one of four counties comprising the "Lowcountry" of South Carolina. The Lowcountry occupies 2,000 square-miles of land at the southern tip of South Carolina. While the extreme wealth concentrated in the Hilton Head resort area skews poverty statistics in this region, poverty is widespread among the GFL initiative's target populations, underscoring stark cultural and racial divisions. The first cluster of farmers, for example, is located on St Helena Island, just off the coast of the city of Beaufort. The participating farmers, nonprofits and the majority of targeted consumers live in Census Tracts 1, 11 and 21. In these tracts, a combined 86 percent of the total population is described as rural. Of the 26,769 people living there, an average 51 percent of the population is African-American, with a *per capita* income of $11,326; and more than 45 percent are listed as "not in the work force" (South Carolina Office of Research and Statistics, 2006).

Project Antecedents and Context

Several economic factors were taken into account in the formation and direction of the GFL project. First, in South Carolina, 92 percent of agricultural businesses are family farms, 97 percent report annual sales under $500,000, and 92 percent cover less than 500 acres (L'Heureux, 2004; U.S. Census Bureau, 2002). By implication, many Beaufort County farms do not support economies of scale nor have sufficient access to capital nor enough technological expertise to compete in the marketplace (i.e., grocery stores, farmers' markets). These farmers generally produce too little to bring in enough revenue to afford needed labor, equipment, or inputs that would make them more competitive. They generate enough profit to live on but not to improve yield, distribution or marketing functions. Some lack knowledge of farm management practices that would reduce their needs to expensive inputs or increase yield inexpensively. Some love to farm but do not like to market their products.

Second, these findings were once again reinforced in 2004 when the LCFB commissioned a study funded by the Donnelley Foundation (Tibbit, 2004). This study concluded that isolated farmers in the county suffered from the absence of cost-effective marketing channels and agricultural management practices. This report also concluded that the local nonprofit feeding programs lack sufficient capacity and training to serve low-income populations in healthy ways.

Despite these obstacles, the Donnelley study also found that the community's rich agricultural traditions and resources, year-round planting season, and the prevalence of farmland owned by low-income populations, offer compelling opportunities for economic development. In particular, the study suggested that these farmers could prosper in growing organic, ornamental and medicinal crops (Tibbit, 2004).

Another economic factor considered was that tourism, South Carolina's number-one industry, is concentrated in this region, producing $9,895,984 in accommodations tax revenues in 2004 – 29 percent of the state's total accommodations tax revenues (South Carolina Commission on Travel & Tourism, 2003). This booming segment supports hundreds of hotels, restaurants and merchants prizing local produce. South Carolina grocers also have joined this trend, providing local farmers with significant retail opportunities (McDermott, 2002).

Finally, in a 2004 state economic development study done for the South Carolina Governor's Office, Harvard economist Michael Porter promoted agricultural clustering to promote competitiveness in South Carolina's rural communities (Porter, 2004). The GFL project aims to encourage these activities at the micro-economic level.

Project Results

In the discussion that follows, we outline how the GFL project is contributing to sustainable agricultural enterprise in a region and how the local economy and emergency food assistance system is being enhanced. We use the three goals of sustainable agriculture as the organizing center for this discussion of results.

Goal 1: Demonstrating Profitability

One major goal of sustainable agriculture movement is that to seek economic profitability while seeking a healthy environment through careful, innovative management practices. In 2005 and 2006, the LCBF and Clemson University entered into a formal arrangement with the owners of eight small-scale farm operations on St. Helena Island in Beaufort County. These farmers committed between .25 acres to four acres of land to growing an agreed upon poundage of vegetables (227,100) for use by the LCFB and its affiliating nonprofits. Farmers could sell vegetables harvested from these acres in local retail markets in response to higher prices and to retain the same level of profits as in 2005. This facet ensured program participation and the maintenance of farm profits. Prices paid to participating farmers (a weighted average of 22 cents per pound across all produce) were based on cost of production values provided by Clemson University's Vegetable Research Center.

The participating farmers and LCFB also formed cooperative efforts in harvesting, storage and distribution activities. A relative large-scale participating farmer agreed to act as a distribution hub but lacked the necessary storage facilities. LCFB paid for building a cement slab and installing a large cooler on their property. Farmers were directly involved in all decision making activities so that the resulting distribution system was workable and seen as equitable by farmers as well as by LCFB.

Farmers indicated an increase in profits resulting from project participation; however, the degree to which profits increased is unclear. Our ability to assess profitability was also clouded because farmers sold 61 percent of the produce from the acreage committed to the GFL project to retail markets. Further, increasing crop yields and quality received more emphasis in this initial year. The eight farmers received a total of $17,367 from LCFB for 82,841 pounds of vegetables, which in turn went to 18 participating agencies for distribution to low-resources individuals.

During the first year of implementation, several challenges arose in trying to balance profitability goals with the food bank's goals of increasing the amount of fresh produce available through the emergency food assistance system.

Balancing Farmer Profitability and Fresh Produce Enhancement

In the first year of operation, project leaders were unsure what conditions and agreements would attract farmer participation, so they were overly generous and required very few commitments that might interfere with each participating farmer's current profit levels. In particular, allowing farmers to sell to local retail markets meant a reduction in local food for the LCFB. For the second year of the project, a more equitable situation will be sought in which the farmers have agreed to deliver to the food bank the promised pounds of produce, in addition to accomplishing their revenue generation goals through retail markets.

A purchase of 350,980 pounds in Year Two and 376,050 pounds in Year Three was envisioned primarily by adding farmers (the original plan was to seek five farmers the first year, eight the second year and 10 total farmers in Year Three). The intent was also to purchase 75 percent of the crop produced off the acreage they committed to GFL project during the first year, 50 percent the second year and 25 percent the third year. From the food bank's perspective the intent was to maintain a steady supply of fresh produce, meeting agency demand levels by increasing the number of farmers affiliated with the initiative. Leaders also assumed that they would be working with farmers who had less access to markets than what the first set of participating farmers actually had.

Realizing Self-Sufficiency and Economic Viability

In order for the initiative to be successful, the food bank ultimately must realize enough revenue from the relationship to continue to pay for the produce received from the farmers. Project leaders thought this would be possible by working with farmers to enhance the quality of their product, the amount of yield realized off their fields and by enhancing the number of markets. The farmers, in turn, must increase their revenues enough to allow them to dedicate a portion of their farm lands to growing produce which will sell to the food bank at below wholesale prices. They also thought that it was possible to secure additional revenues by acting as a distributor and marketer of at least some of the farmers' produce. Self-sufficient funding of the GFL initiative remains a challenge for LCFB, but enough of what was envisioned is happening that project leaders will continue to develop their original strategy. While the LCFB received substantial grant funding, the ultimate goal is to eliminate grants as a means of support. Current sustainable revenue streams (covering 75 percent of cost) include payment of shared maintenance fees ($8,284). The LCFB also charges farmers a modest fee for trucking produce to market and acting as their broker at local farmers' markets and grocery stores. Because grant funds currently cover the cost of buying food from local farmers, revenues are devoted to hiring additional drivers and capital investments (in particular, no interest loans for farmers) in the participating nonprofits and farmers.

Goal 2: Achieving and Sustaining Health by Increasing Access to Nutritious Food

A second goal of those seeking sustainable agricultural situations is that they strive to use natural resources in a way to promote people's ability to achieve or sustain health. During the late 1990s, less food was available to food banks across America than during previous periods. The plight of the hungry is more acute now than in the previous decade (America's Second Harvest, 2006). Fortunately, the LCFB has been able to increase the food it has available by creative local efforts, including the GFL initiative.

A total of 757,420 pounds of GFL produce and additional food and related household items were distributed through the 18 participating agencies during Year One, a 40 percent increase over the prior year when 549,344 pounds were distributed[2]. Based on LCFB inventory tracking records concerning the quantity and types of food distributed to the participating agencies, in the first year of the GFL project implementation, 82,841 pounds of vegetables produced locally were purchased by LCFB from local small farmers, and 64,130 pounds were distributed to the collaborating agencies, which accounted for eight percent of the total food and related items received from the LCFB[3,4]. The remaining 20,711 pounds were distributed to participating faith-based coalitions when there was an overabundance of crop, more than what the participating agencies could use. In addition, the LCFB distributed 1,558 pounds of fresh vegetables obtained from other sources. During the previous year, no fresh vegetables were distributed to the agencies.

Low-resourced individuals lack access to an adequate supply of nutritious food on a consistent basis. In particular, the emergency food system often does not provide access to nutritious food. In fact, there is some scientific evidence linking food insecurity and obesity which in turn raises the risk of heart disease, diabetes, high cholesterol levels, among other health conditions (Basioitis and Lino, 2003; National Institute of Health, 1998). The inclusion of fresh fruits and vegetables available to families receiving food from the LCFB agencies is a practical and important means for optimizing nutrition to reduce disease risk and maximize good health (Van Duyn and Pivonka, 2000).

Goal 3: Equitable Access to Markets and Other Resources

A third major goal of sustainable agriculture initiatives is that they strive to address or correct, as well as do no further harm to, the social conditions and seek equitable access by all to market, environmental and community resources. Sustainable agriculture involves consideration of the local cultural attitudes including farming methods and farmer goals to include life style choices. In our case, the local attitudes about standards of practice in the emergency food assistance system are also important.

Eight older men and one woman (spouse/co-owner) comprised the first group of committed farmers. All are in their mid-fifties or older, so succession is an ultimate concern at least for GFL leadership. Three farmers rent their land from local families.

Farmer income is supplemental for two farmers who are employed elsewhere, although one plans to retire into farming. Two senior farmers have farming as their primary source of income.

GFL participants use practices typical for local small-scale farm owners and face typical problems including depleted soil from ineffective cropping practices, water management issues, and high salinity due to the coastal location. Farm chemical use varies between over and under use by farm and field, but alternatives to pesticides have not been explored. Ineffective irrigation systems lead to water loss in some cases while poor ditching results in flooding in other cases. The quality of crops varies greatly among farms although this problem had greatly improved by the end of the first year. Crop management systems that impair soil quality often result in greater inputs of water, nutrients, pesticides, and/or energy for tillage to maintain yields. All but one of these farmers need to improve crop management practices to gain greater efficiency.

Based on soil and water samples farmers were coached concerning improved economically sound irrigation and water disposal, soil management, and pest prevention methods resulting in some improvement in yield levels and quality. While some improvements have occurred, dealing with water issues may be necessary to achieve sustainability. For example, only three GFL farmers have adopted recommendations concerning drip irrigation. While individual consultation will continue to occur, demonstrations will be done on one or more of the farmer's fields as a way to enhance management practices. The important issues related to water quality involved salinization and contamination of ground and surface waters by pesticides, nitrates and selenium.

Even though the LCFB staff has noticed an increase in consistency of product delivered by the farmers, room for improvement remains. Farmers are being encouraged to continue to improve practices to subsequently improve product quality and enhance yields.

Having the farmers agree to grow certain produce has created a more diverse cropping practice for some of the farmers. This should help to create a more sustained ecology over time and there is some evidence that it has helped suppress weeds, pathogens and pests. Diversification is also being encouraged as a way to partially control deer that are foraging on one of the farmer's fields, although nothing has been found yet to control the problem completely. The goal of GFL is to develop efficient, biological systems which do not need high levels of material inputs.

With the affiliating nonprofits, discussions have occurred about what produce they would like to receive including increasing the amount of nutritious food made available to their consumers. Agency leaders also discussed the value of eating fresh produce, how to store it, handle it and cook it in healthy ways.

The LCFB and Clemson leaders have worked through the development, piloting and revisions of all aspects of the project resulting in a fundamental transformation of operations. The entire operation has been reviewed, revised, and in some cases augmented. New management systems, inventorying procedures, and service functions have been added. There has also been a conscious effort to experiment with procedures and activities that will lead to a sustainable system that is not reliant on grant money to sustain the operation.

Other impacts have occurred since the beginning of the project. Some of the more important ones are reviewed in the next section.

Local Economic Impact

Many authors (Goldschmidt, 1946; Haweil, 2002; and Pirog et al., 2001, among others) include local economic vitality in their definition of sustainable agriculture. Local food advocates generally see local economic growth resulting from import substitution, which is the replacement of products purchased outside the local economy with local production (Homm et al., 1991). The rationale for import substitution is based on economic multiplier analysis, where the level of local purchases in term determines further buying in the local economy (Hughes, 2003).

The IMPLAN input-output model of Beaufort County yielded economic multipliers for vegetable production in the County. According to LCFB data, four jobs were expected to be generated or at least retained by the GFL project. The job multiplier (total or multiplier based jobs divided by direct farming jobs in this case) for vegetable and melon farming in Beaufort County was estimated to be 1.57. Based on this multiplier, the total current impact of the farming based portion of the GFL project on the local economy is 6.3 jobs. While not a large number, this estimate represents employment generated for farmers and others who usually have lower than average incomes compared to national averages in an area where economic opportunity is limited. Hence, the GFL program is starting to make a contribution to poverty alleviation and economic growth in Beaufort County.

Sustainability and Food Transportation Cost

A significant portion of the potential societal benefits of local food systems arguably arises from transportation cost savings. Local food advocates point to the reductions in energy consumption and environmental pollution from meeting food needs with local products. On the other hand, other literature cites the decrease in real (accounting for inflation) transportation cost as a key long term historic trend. This latter viewpoint at least implies that transportation cost savings from local foods versus other sources does not provide substantial societal benefits. One can also draw the inference that production centers based elsewhere (where economies of scale in production and industry cluster or agglomeration benefits may hold) will continue to hold cost advantages over local food systems.

In our view, the debate centers on the adequacy of market prices to reflect true social costs and benefits, especially in terms of environmental degradation and resource scarcity, in particular in terms of petroleum consumption. Another wildcard is the role of government infrastructure development as a subsidy for transportation systems.

Priog et al. (2001) argue that both the value of reduced emissions, and hence reduced contributions to greenhouse gases, as well as the value of saved fuel and other relevant transportation cost savings should be included in comparisons of local versus typical food distribution systems.

Discussions with GFL farmers and a pre-production report indicated that squash, tomatoes, and collards tended to be important crops. However, discussions with LCFB officials indicated that locally produced vegetables supplemented Lowcountry LCFB supplies rather than substituting for production procured elsewhere. Hence, the transportation cost savings touted by local food advocates are as yet unrealized in this situation. However, it is still instructive to evaluate the potential impacts of enhanced local food production as demand increases for vegetables by LCFB users.

Evaluation of daily count data for all of 2006 for terminal markets in Columbia, South Carolina, and Atlanta, Georgia (collected by the U.S. Department of Agriculture Agricultural Marketing Service, 2007) indicated that crops produced by Lowcountry small-scale farmers could serve as a substitute for production coming from at least fairly distant areas in many cases. For example, squash from Florida was the primary point of origin for shipments in the Columbia Market (at 29.1%) followed by South Carolina (20.8%), Georgia (13.9%), Mexico, North Carolina, and Michigan. Also as expected, shipments followed a seasonal pattern. For example, shipments from Mexico to the Atlanta Market (consisting of 26.2% of counted shipments) were heaviest starting in November, going through April and then declining to low levels in the late spring and summer months. The bottom line is that local foods sold by local farmers to LCFB could one day provide savings in terms of reduced energy consumption and lowered greenhouse emissions

Transformation of the LCFB Operations and Mission

In the case of the LCFB, during the first year of the GFL initiative, management systems for purchasing and tracking product were enhanced. Staff and affiliating agency reporting procedures were enhanced. The entire inventory was assessed and rated for its nutritional value. Personnel changes were made to streamline all aspects of the operation and to add on a nutrition educator. The accounting and inventory systems was enhanced. Agency relations were enhanced. Food safety inspections and training were enhanced. The services offered by the food bank were expanded. In essence, the operation, services, and outputs were improved during the first year of GFL initiative.

Despite efforts to increase the volume of fresh produce and to create a storage system that would allow agencies to get the volume of produce they could effectively handle, three agencies stopped participating because they could not handle the volume of produce received. GFL leadership will continue to develop effective systems for helping nonprofits handle higher volumes of fresh produce, as well as developing ways for the LCFB to handle larger volumes of fresh produce so that it is distributed to the needy. Leadership also plans to expand the use of mass distribution sites by the development of cooperatives so that additional value can be added to this effort and more sources of revenue created to sustain the effort.

The LCFB is the single most important source of food and related household items for the participating agencies. The year prior to the GFL project implementation, on average 75 percent of their food and related household items was obtained from the LCFB. At the end of the first year of GFL, 79 percent on average was obtained from LCFB. Given that the quantity of food and related items received from the LCFB has increased by around 40 percent, the five percent increase in the percentage of LCFB food used in the agencies' programs indicates that the agencies have also been able to obtain more donations from other sources, although this increase has been lower than the increase in the amount of food and related items received from the LCFB.

Summary and Conclusions

Traditionally food banks are seen as a major player in America's emergency food assistance system. They are not seen as a contributor to community economic development in an area, other than very indirectly. The initial leadership of the LCFB had a different vision.

Our analysis reveals that during the first year, the link between sustainable agriculture and community development was made clear as the LCFB provided delivery, storage, inspection, and disposal services to farmers and affiliating nonprofits. It became a micro-lender to small scale farmers. It distributed produce for small scale farmers not only to agencies affiliated with the emergency food assistance system but also to retail markets. It inspected facilities to enhance food safety. It sought additional capital on behalf of agencies and farmers that lacked capacity to seek such capital. It engaged in value-added efforts by taking produce from farmers and combining that produce with additional items so the items received by affiliating agencies were now more nutritious. As a result, local farm incomes were apparently enhanced albeit to an unknown degree.

The project also served as a capital lender to 18 participating nonprofit agencies. Perhaps more importantly, the operation of the LCFB itself was transformed to a more sustainable business model through the implementation of such activities as an inventory tracking system.

Before the project began, very little fresh produce was consistently distributed through the food bank system. Other items such as juices and cereal were high in sugar content. After one year of GFL operation the food bank inventory looks very different. More food items that have great nutritional value are present. While not the total answer, altering the way food assistance providers do their work and providing greater access to nutritious food are key factors to creating positive health outcomes for low-resourced individuals in the Lowcountry.

These outcomes suggest that the LCFB and its GFL partnership could provide a model for linking sustainable agriculture practices with innovative community development strategies that stimulate new small-scale enterprises and create new food distribution and marketing channels.

NOTES

1 Dr. Kathleen K. Robinson, Research Professor, Institute on Family and Neighborhood Life, Clemson University and Mr. Robert Strickland, Director of Business Operations, Lowcountry Food Bank are Co-Directors of the Growing Foods Locally Project. The project is currently funded by grants from the USDA, Sisters of Charity Foundation of South Carolina, The Donnelly Foundation, and the Mary Reynolds Babcock Foundation.

2 Much of the information concerning local food systems is presented under the topic of direct farmer marketing, where farmers make sales directly to final consumers or retail outlets. While there is a large amount of overlap between the two concepts, they are not the same. For example, direct market includes internet sales, where customers are usually not local. Local food systems include community gardening, which does not involve commercial farmers. Even farmers' markets are not entirely local foods, because some farmers' markets have tourist as a major segment of their market.

3 These reports come from LCFB staff who weight food purchased by each agency. These reports indicate the kind of food requested by agencies. The summaries of the inventory records for the twelve months before (October/04 - September/05) and after (October/05 - September/06) the GFL project started were reviewed.

4 Monthly meal reports that participating agencies must file with the Food Bank are another possible data source. However, some agencies did not file reports, others had staff lacking the math skills to calculate necessary percentages, and incomplete reports indicated that others did not devote required time and effort to record keeping including properly weighing food. To increase accuracy in reporting, LCFB's Agency Relations staff has conducted several training sessions with agency personnel to enhance accuracy, but only with mixed results. Because these reports contact a variety of potentially useful data, the hope is to combine this data source with Food Bank inventory records to gain a detailed evaluation of the GFL project's impact on the nutritional content of distributed foods.

REFERENCES

Alward, G., E. Siverts, O. Olson, J. Wagner, O. Senf and S. Lindall. (1989). MicroImplan Users Manual. St. Paul, MN: University of Minnesota, Dept. of Agricultural and Applied Economics.

Barkley, D. and P. Wilson. (1995). The Role of Nontraditional Agriculture in Rural Development. University of Wisconsin-Madison, Department of Agricultural Economics, Community Economics Newletter (221), March.---.

1992. Is Alternative Agriculture a Viable Rural Development Strategy Growth and Change. 17(2): 239-253.

Basiotis, P.P. and M. Lino. 2003. Food Insufficiency and Prevalence of Overweight among Adult Women. *Family Economics and Nutrition Review*, Vol. 15, 55-57.

Brown, C. Consumers' Preferences for Locally Produced Food: A Study in Southeast Missouri. *American J. of Alternative Agriculture* 18(2003): 213-224.

Cohen, R., J. Ohis, and M. Kim. (2006). Hunger in America 2006: Final Local Report Prepared for The Lowcountry Food Bank (4101). Princeton, NJ: Mathematica Policy Research, Inc.

Duffy, P. A.; G. G. Hallmark; J. J.; L. Claxton; C. Bailey; and S. Mikloucich. (2002). Food Security of Low-Income Single Parents in East Alabama: Use of Private and Public Programs in the Age of Welfare Reform. *Rural Sociology*, 18(1), 48 – 81.

FPC. *Attracting Consumers with Locally Grown Products*. Lincoln, NE: Food Processing Center, Institute of Agriculture and Natural Resources, University of Nebraska. 2001.

Goldschmidt, W. J. (1946). Small Business and the Community: A Study of the Central Valley of

California and the Effects of Scale of Farm Operation. Washington, DC: U.S. Senate.

Harrison, B. (1992). Industrial Districts: Old Wine in New Bottles? *Regional Studies*, 26(5),469-483.

Haweil, B. (2002). Home Grown: The Case for Local Foods in a Global Market. In T. Prugh (Ed.), *Worldwatch Paper 163*. Danvers, MA: Worldwatch Institute.

Hilchey, D, T. Lyson, and G. W. Gillespie Jr. (1995). Farmers' Markets and Rural Economic Development. *Community Agriculture Development Series,* Ithaca, NY: Cornell University, Farming Alternatives Program, Department of Rural Sociology.

Homm, D., G. Doeksen, S. Ralstin, M. Woods, and J. Frye. (1991). Blueprint for Your Communities Future: A Trageted Industry Approach.F-918, Stillwater, OK: Oklahoma Cooperative Extension Service.

Hughes, D. W. (2003). Policy Uses of Economic Multiplier and Impact Analysis. *Choices, 18(2),* 25-30.

Ikerd, J. (2002). Revitalizing Rural Communities Through Agriculture. Paper presented at *Alliance for Recapturing Wealth on the Canadian Prairies*. Conference sponsored by Agricultural Renewal Alliance, Government of Manitoba. Brandon, Manitoba, October 24th-25th.

Knight Ridder/TribuneBusinessServices b. (2004). "The State" Newspaper, *Rural Areas Want Role In Clusters*, July 18, 2004, pg. F1. Leatherman, J.C. and D.W.Marcouiller. (1999). Moving Beyond the Modeling of Regional Growth: A Study of How Income is Distributed to Rural Households. *Economic Development Quarterly*. 13(1),38-45.

L'Heureux, D. (2004). Family Farms Remain Predominant in South Carolina Agriculture. *The State* (Columbia, SC), February 4, 2004.

Lowcountry Food Bank. (2006). http://www.lowcountryfoodbank.org/index.asp.

Lyson, T. A. (2004). *Civic Agriculture: Reconnecting Farm, Food, and Community* Medford, MA: Tufts University Press.

Lyson, T. A., G. W. Gillespie Jr., and D. Hilchey. (1995). Farmers' markets and the local community: Bridging the formal and informal economy. *American Journal of Alternative Agriculture* 10:108-113.

McDermott, J. P. (2002). "The Post and Courier" (Charleston, South Carolina), *South Carolina Farming Industry Profits from 'Organic' Movement*, April 29.

Molnar, J. J., P. Duffy, L. Claxton, and C. Bailey. (2001). Private Food Assistance in a Small Metropolitan Area: Urban Resources and Rural Needs. *Journal of Sociology and Social Welfare*, September, Vol. 28, No. 3, pp. 187 – 209.

National Institutes of Health. (1998). Clinical Guidelines on the Identification, Evaluation, and Treatment of Overweight and Obesity in Adults: The Evidence Report. http://www.nhlbi.nih.gov/guidelines/obesity/ob_gdlns.pdf

Piore, M. J. , and C. F. Sabel. (1984). *The Second Industrial Divide*. New York, NY: Basic Books, Inc.

Pirog, R., T. Van Pelt, K. Enshayan, and E. Cook. (2001). "Food, Fuel, and Freeways: An Iowa Perspective on How Far Food Travels, Fuel Usage, and Greenhouse Gas Emissions." Leopold Center for Sustainable Agriculture, Iowa State University, Ames. July.

Porter, M.E. (1998). Clusters and the New Economics of Competition. *Harvard Business Review*. November-December. ---.

(1985). *Competitive Advantage: Creating and Sustaining Superior Performance*. New York: The Free Press.---.

(2004). *Competitiveness in Rural Regions: Learning and Research Agenda.* Boston: Harvard University, Institute for Strategy and Competitiveness, Harvard School of Business, February.---.

(2000). Location, Clusters, and Company Strategy, in G. Clark, M. Gertler, and M. Feldman, (Eds). Oxford Handbook of Economic Geography, Oxford: Oxford University Press.---.

(2000a). Location, Competition and Economic Development: Local Clusters in a Global Economy. Economic Development Quarterly, Vol. 14, No. 1, February, 15-34.

Porter, M. E. and C. van der Linde. (1995a). "Green and Competitive: Ending the Stalemate" *Harvard Business Review*, September-October.---.

(1995b). Toward a New Conception of the Environment-Competitiveness Relationship. *The Journal of Economic Perspectives*, Vol. 9, No. 4, Fall.

Second Harvest. (2006). www.secondharvest.org.

Smith, P. K. and S. L. Hoerr, (1992). Comparison of Current Food Bank Users, Non-Users and Past Users in a Population of Low-Income Single Mothers. *Journal of Nutrition Education*, Vol. 24, 59 – 65.

South Carolina Commission on Travel and Tourism. (2003). Leisure and Hospitality Industry Employment in South Carolina 2003. Columbia, SC. www.discoversouthcarolina.com.

South Carolina Office of Research & Statistics. (2006). South Carolina Community Profiles by County, Beaufort County Census Tracts 1 and 11, www.sccommunityprofiles.org.

Sterns, J. A., D. B. Schweikhardt, and H. C. Peterson. (1998). Using Case Studies as an Approach for Conducting Agribusiness Research. *International Food and Agribusiness Management Review*, 1(3), 311-327.

Strickland, Robert C. (2006). Correspondence of the Lowcountry Food Bank Director of Operations. November 12.

Tibbit, C. (2004). The Feasibility of Working with Small Scale Farms to Enhance the Supply of Nutritious Food to the Lowcountry Food Bank. A consultant report delivered to the Director of the Lowcountry Food Bank in partial fulfillment of a Donnelly Foundation grant.

University of California Sustainable Agricultural Research and Education Program. (2007). http://www.sarep.ucdavis.edu/

Van Duyn, M.A.S. and E. Pivonka. (2000). Overview of the Health Benefits of Fruit and Vegetable Consumption for the Dietetics Professional. *Journal of the American Dietetic Association*, 100(12):1511-1521.

Options for the Economic Health of Farmers, Farmers Markets and Communities: Homebased Fruit and Vegetable Microprocessing

Sandra Bastin

Kentucky passed farmer friendly legislation in 2003 allowing farmers to process value added food products in their home kitchens with the emphasis on community economic development and farm to table concepts. A unique partnership between the University of Kentucky Cooperative Extension Service, Kentucky Cabinet for Health Services, Kentucky Department of Agriculture, and Farm Bureau allows farmers to sell home canned foods from the farm, farmers market, or certified roadside stand. The value-added products must contain a fruit, vegetable, or nut grown by the Kentucky processor. Farmers processing low risk products simply register. Over 289 farms in Kentucky are selling items such as pumpkin bread, blueberry jam, and pear preserves. To produce low acid or acidified products, farmers must attend a workshop, have recipes approved by a process authority, develop a label, and make application for a fee. Over 200 farmers have been trained with 36 farmers producing diverse products such as salsa, pickled squash, and corn relish. These products allow farmers to add economic value to their raw produce, reduce waste, and use blemished but otherwise quality produce. Extending the selling season allows the opportunity to strengthen the farm to table link within the community. Without the large capital outlay of a permitted kitchen, farmers can also find which products may be successful for commercial market venues. Data indicates farmers are promoting these concepts within their communities by going beyond traditional marketing of fresh produce and participating in homebased microprocessing to promote value-added products.

Historically, tobacco has been the foundation for many of rural Kentucky's farm profits. However legal and political challenges have prompted farmers to explore avenues of broadening their market opportunities. Across Kentucky, farmers are searching for new ways to sustain their rural way of life. One such opportunity is House Bill 391 (HB 391), a farmer friendly legislation that enables farmers to produce safe value-added fruit or vegetable products from their home and then sell from farmers markets, their farm or Kentucky Farm Bureau Certified Roadside Markets. This new dimension for local food systems gives local producers an increased capacity beyond just picking and selling. Since there is a lack of literature on how direct marketing truly impacts the production systems and economic health of farmers and rural communities, programs such as homebased microprocessing will help set the stage toward successful direct marketing strategies (Brown, 2002).

Sustainable Agriculture

According to United State Department of Agriculture Sustainable Agriculture Research and Education (USDA SARE, 2007) the primary goals of sustainable agriculture include providing a more profitable farm income, promoting environmental stewardship, and promoting stable, prosperous farm families and communities. This move to sustainability creates a direct economic and social partnership between farmers and local community members, strengthening the connections between the community individual and the farmer and their land. The recipe for successful sustainable agriculture in Kentucky includes economic profitability, environmental stewardship, and social responsibility (Ikerd, 2000).

Feenstra and Campbell (1996-97) describe a local food system as "a collaborative effort to integrate agricultural production with food distribution to enhance the economic, environmental, and social well-being of a particular place (i.e.: neighborhood, city, county or region)." Peters (1997) states, "One of the primary assumptions underlying the sustainable diet concept is that foods are produced, processed, and distributed as locally as possible. This approach supports a food system the preserves local farmland and fosters community economic viability, requires less energy for transportation, and offers consumers the freshest foods." This view is becoming the norm in thinking about food systems, both within local communities and in educational and government institutions (Dahlberg, 1999; Hinrichs, 2000; Renting, et al., 2003; Thomson, et al., 2006). Interest in food systems will continue to grow as researchers delve more deeply into the social, economic and environmental issues surrounding food systems.

Kentucky has identified farmers markets, the creation of value-added enterprises, and direct marketing as a means of diversification leading to agriculture sustainability (Kentucky Department of Agriculture, 1996). Direct marketing of farm products has seen a rise in recent years (USDA AMS, 2004), supporting Kentucky's focus. New enterprises include cooperatives, agritourism, and homebased businesses. While these are not novel concepts, they are new to the food arena. Farmers are promoting these concepts in their communities, by going beyond traditional marketing of fresh produce and participating in homebased microprocessing to promote value-added products at their local farmers markets.

Farmers Markets

Farmers markets are becoming an integral part of the urban/farm linkage and are rising in popularity. Farmers supplement farm income by selling directly to consumers. Consumers have the opportunity to interact with the farmer who grows the produce and access to locally-grown produce, and the community gains easy access to food while enjoying the promotion of nutrition education, wholesome eating habits, and better food preparation, while boosting the community's economy (USDA AMS, 2002). To support the concept of sustainable agriculture and local foods systems, Kentucky passed the previously-discussed HB 391. Coupled with training, HB 391 provides farmers with a viable option in the selling of safe, quality value added foods at farmers markets.

Farmers markets provide two key roles in the community. Farmers provide fresh, safe, high-quality foods and provide an opportunity for the community to interact with farmers and learn more about their way of life (Wilkins & Eames-Sheavly, 1999). The visit to the farmers market is all about the experience. The emphasis on more fruits and vegetables in the diet is just another way to link consumers with the farm. Because many farmers are able to share a special quality and pride in what they grow, showing the benefits of supporting local farms, consumers are drawn back to the farmers market. The tie-in with local restaurants, schools, and supermarkets can strengthen their position with the community.

There are many channels available to market farm products. These markets are designed to offer quality products and "on-farm" experiences. They include greenhouses, garden centers, agriculture tourism, festivals, pick-your-own, wholesale, roadside stands, community supported agriculture, and farmers markets. The trend to visit farmers markets is on the rise in Kentucky (Ernst & Woods, 2006). It seems to be fueled by consumer interest, the farmers desire to make more money, and growing community awareness of the contributions of farming and local food systems. Farmers markets are perfect for small farm producers who are either uninterested or unable to participate or compete effectively in larger wholesale channels. While farmers markets are still a small percentage of total agriculture sales, the number of farmers participating in farmers markets are growing (Table 1). As farmers become more creative in offering more diverse value added products, markets will become more economically feasibility and contribute to the economic development of the community.

Table 1. Kentucky Farmer Market Growth and Interest in Selling Value-Added Foods

Year	Number of Markets & Percentage Increase	Gross Sales	Number of Vendors	Markets Wanting to Offer Value Added Products & Percentage of All Markets
2004	96	$5 million	1548	27 (28%)
2005	98	$6 million	1678	40 (42%)
2006	107 (13%)	$7 million	1808	57 (53%)

The number of markets has been steadily increasing since the Kentucky Department of Agriculture began working with farmer market managers and vendors (Eaton, 2006). With the formation of the Kentucky Farmers Market Association, it is expected that growth will continue and farmers markets will become more viable as the association takes on an expanding role in their success. The new implementation of Farmers Market Summits will strengthen leadership at the local level and provide an avenue for education. The Farmers Market Summits are two-day, regionally held conferences where farmers and interested parties can participate in workshops about market rules, food safety and sanitation, cooking demonstrations, product expansion, marketing, promotional campaigns, and voucher expansion with Women, Infant, and Children (WIC) and seniors, as well as network with other attendees. These activities, plus a new *Kentucky Farmers Market Manual*, published jointly with the Kentucky Department for Public Health, will help provide cutting edge information and technology for innovative thinking and action at Kentucky's Farmers Markets.

The Case Study

In 2002 several vendors at the Woodford County Twilight Festival were required to cease operation by health department officials. At this time it was determined that the vendors were not in compliance with state laws requiring that their products be produced under a "commercial" food manufacturing permit. Domestic home-type kitchens do not meet the state requirements for a "commercial" permit.

For five years, Imogene Antrobus had been selling her pickles, relishes, and salsa at the local farmers market. Her goal was to bring home a $3,000 profit, allowing her to make ends meet since tobacco sales opportunities had been reduced on her family farm. She asked the County Cooperative Extension Service Horticulture Agent to research how other states regulated home-processed products. She found there was a trend to allow jams and jellies to be made in the home, but none allowed for home processing of low acid or acidified food products. The trend allowed anyone with the appropriate permit to produce low risk items, but did not necessarily require home grown produce to be a part of the recipe. The ensuing fallout led to a change in Kentucky law that allows farmers, with proper training, to process their homegrown fruits and vegetables in their own kitchen and then sell them from their farm or at the farmers market. HB 391 and Farmers Market legislation, along with a large group of implementation partners, have infused Kentucky agriculture with a new entrepreneurial spirit.

Collaborative Efforts

What really made the vision come to a reality was the hard work and foresight of a group from a variety of backgrounds. The broad representation of the group brought different perspectives and needs, providing a big picture that ultimately led to the success of the program. HB 391 was a collaborative effort between farmers, legislators, the University of Kentucky Cooperative Extension Service, Food Safety Branch of the Kentucky Cabinet for Health Services, Kentucky Department of Agriculture, and Kentucky Farm Bureau (Bastin, 2006).

As with any group, building consensus was the most difficult part of meeting our goals. Typically, building consensus is a process involving meeting, clarifying responsibilities, deliberating, deciding, and then implementing agreements (Susskind, et al, 1999). My role was to listen and mediate, often summarizing in a non-threatening way to keep all parties on track, during the meetings. The eight month process involved much time and effort by all parties, and was not without some level of discomfort by all. In any group, individual parties are interested in their own piece of the pie. However, the goals were worthy and the timing was right (since two legislators were interested in supporting farmers) for us to all agree to write legislation that allow farmers additional opportunities for economic profits, while strengthening community and farm to table concepts.

This unique cooperative effort was an opportunity to meet the three goals of homebased microprocessing as determined by the partners involved:

To showcase a number of genuine home grown, value added food products at select locations.

It provides a distinctive niche for farmers to share their proud Kentucky heritage of producing fine quality food products. Local citizenry and Kentucky tourists alike now have the opportunity to acquire home processed food items long treasured by friends and neighbors of the Kentucky farmer; value added products derived from locally grown, harvested, and processed foods.

To allow farmers to provide safe, quality value added products to their communities while capturing more of the retail food dollar and strengthening the local economy.

The addition of an array of diverse products to the table has drawn more Kentuckians to these outlets to shop, while extending the selling season, and broadening the consumer base. The strength of these markets will be enhanced and contribute to increased economic health in many rural Kentucky communities, while promoting local value added foods and improving the farm to table connection.

To provide research based methods and recommendations for farmers to produce safe, wholesome, value added foods.

Because of particular health risks associated with processed foods, education is a key component and is designed to enhance the homebased processor's awareness of relative food safety risks and to arm the homebased processor with current good manufacturing practices. When food sanitation and safety guidelines are faithfully implemented during processing, the goal of a safe, wholesome product will be realized. The University of Kentucky Cooperative Extension Service and Cabinet for Health Services Food Safety Branch, by providing education, is also encouraging consistency among participants and inspectors for a better experience for consumers.

The implementation process included the writing and sponsorship of HB 391 and the development of training materials to support a Homebased Microprocessor Workshop. The workshop would provide training on the law and proper safety, sanitation and processing techniques to ensure the safety of consumers purchasing the value added products.

Homebased Processing and Microprocessing Program

The farm friendly collaborative efforts which culminated in the passaged of HB 391 and Famers Market legislation have provided an opportunity for Kentuckians to sell homebased products in certain locations, if the final product contains a primary or predominant Kentucky-grown ingredient that was grown, harvested, and processed by the farmer. These laws allow the home preparation of foods that do not pose a significant health risk to the buying public. It also calls for in-state training and certification for home microprocessors of low acid or acidified foods that may pose health risks if processed incorrectly. For eligible farmers, HB 391 waives the former requirements that home processed foods must be made in a certified or permitted kitchen that is separate from the home's regular kitchen. Farmers currently using a permitted kitchen to process food may continue to do so.

Currently, two categories are available for farmer participation: homebased processor and homebased microprocessor. Both require that the value added products processed must contain a fruit, vegetable, or nut grown by the processor. The farmer must grow, harvest, and process the product for sale. The products may only be sold from the processor's farm, a Farm Bureau Certified Roadside Stand or Farmers Markets listed with the Kentucky Department of Agriculture. Both categories must label products as indicated by the law and limit annual sales to less than $35,000 for each product sold. Farmers who find sales more than $35,000 will then need to take the next of obtaining a commercial food manufacturing permit and using a permitted kitchen to process their product. A commercial food manufacturing permit allows the sale of value added products at additional locations such as restaurants, grocery stores, gift shops, and through the internet.

No fee is associated with becoming a homebased processor. An annual application form must be filed with the Kentucky Cabinet for Health Services Food Safety Branch. Homebased processors may produce and sell any of the following lower risk products: whole fruits and vegetables; mixed greens; dried herbs, nuts, fruits and vegetables; jams, jellies, preserves and sorghum; and bread, cakes, pies and cookies. The key is that the value added products processed must contain a fruit, vegetable, or nut grown by the Kentucky processor and the sales locations are limited. Currently, there are 289 homebased processors registered across the state. This is a 335 percent increase over the first registered homebased processors in 2003.

Homebased microprocessors are permitted to sell higher-risk, shelf stable low acid or acidified foods, including salsa, barbecue sauce, herbal vinegars, pickled fruits and vegetables, and pressure canned vegetables. For a fee of $50, homebased microprocessors must attend and successfully complete the University of Kentucky Homebased Microprocessor Workshop. This workshop is designed after the Food and Drug Administration's Better Process Control School which is required for supervisors of food manufacturing companies. Presented by myself and a Food Safety Branch representative, the workshop includes information on the law, food safety and sanitation, home canning, and the microbiology of low acid and acidified food products. Participants must successfully complete two exams that are administered during the workshop, to receive a certificate which must be updated every three years. Cooperative Extension Agents administer a video and exam to farmers for certification extension. It is too soon to see the percentage of recertifications that will occur in the program.

In addition, each food product must have a standardized recipe approved by a process authority. To date, individual process schedules approved includes 268, with only four not approved. The variety is limited only by their imagination as farmers try to provide customers with that 'just picked" flavor, "just like Grandma" used to make. These marketing ploys have proved to be a big attraction, especially for urban consumers at local farmers markets who may remember when their grandma gardened or canned at home. Farmers participating in the homebased microprocessing program usually take two to three years to determine what products they are going to sel!. As they begin to sell products, consumer requests drive additional recipes to be produced. Product labels must be approved as well. Upon completion of the workshop and approval of recipes, an application with a fee of $50.00 must be filed annually with the Kentucky Cabinet for Health Services Food Safety Branch. Food Manufacturing Inspectors associated with the Food Safety Branch may inspect the home kitchen initially and consequently every two years. A mechanism is in place to handle complaints or non compliant participants.

A total of 81 agency professionals and 193 farmers received technical expertise and knowledge from 22 Homebased Microprocessor Workshops. The Value Added Microprocessor Project (SARE Project ES04-072) supported the development and implementation of materials used for training in these workshops. The workshops were presented in a variety of locations so everyone had access to the training. Four workshops per year are offered and locations are based upon demand. The following Kentucky map compiled by the Food Safety Branch shows where the current 36 homebased microprocessors are selling their value added products (Figure 1). This is a 500 percent increase over the previous year's registration. Because of the requests from farmers, this number is expected to continue to rise.

Figure 1. Current Location of Homebased Microprocessors in Kentucky

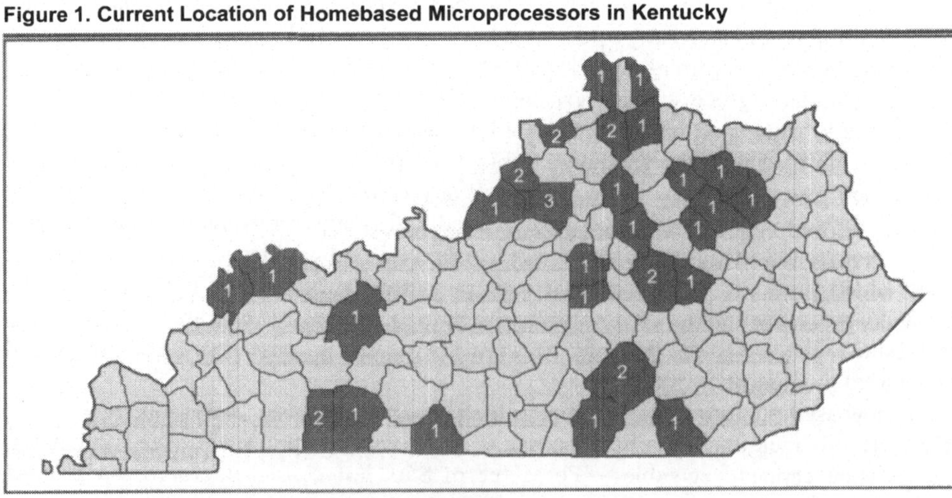

During the workshop, participants take a pre-test and two exams after material is presented. Data analyzed from exams indicated there is a large increase in knowledge among participants. All of farmers indicated they will follow safe processing procedures according to approved process schedules and legislative rules. Spot check compliance visits have found all registered homebased microprocessors to be following the regulations of HB 391. Compliance visits also identify unregistered sellers and provide the opportunity to educate these farmers.

Success Stories

Careful business planning and marketing are important for any successful business venture, and it is no different for Kentucky's homebased microprocessors. The following individuals attended the homebased microprocessing workshop, wrote a business plan, and considered a variety of marketing schemes. Each knew they had customers willing to buy their product and the farmers market offered them the outlet to sell those products, if they were able to maintain quality and be competitively priced. With a little ingenuity, a growth in sales allowed the addition of other value-added products, increasing their profit margins. Their stories come from personal interviews.

- Linda Ison, owner of Crosswind Llama and Herb Farm in Crestwood saw her profits double the past two years. It's not because she selling more llamas and herbs, the mainstay of her 16-year-old farm. Linda attributes her success to herbed French bread and home canned hot pepper jam and asparagus spears. These are just a few of the 18 value-added products she's selling at her local farmers market as a result of HB 391 and being trained as a homebased microprocessor. The farmers market where Linda sells averages $13,650 per vendor. The market enjoys an excellent urban location and community support.
- Bobby Giles, owner of Honey Tree Farm has expanded his product line from honey to include 15 microprocessed value-added products, mostly using fruits in jams and jellies and cucumbers in pickles and relishes. In addition to the University of Kentucky Homebased Microprocessor Workshop, he also attended From Kitchen to Market, hoping to gain the entrepreneurial knowledge needed to successfully launch a full commercial product line. Bobby listens to his customers and adds new products each year to meet the demand. In his community, Bobby has encouraged other farmers to participate in the homebased microprocessor program.
- Among the first trained microprocessors were the McDowell sisters. Perfecting their grandmother's recipe, they began selling salsa across the state at every farmers market and festival they could. After the sale of 10,000 jars at $5.00 per pint, they were awarded a Kentucky Board of Agriculture grant for $50,000 to renovate their basement into the permitted kitchen needed to sell commercially. After success in regional markets, they are currently pursuing an agreement with Walmart.

These are just a few of the examples of how this program has allowed for increased profitability. In addition, these farmers are having fun selling food products they are proud of; they're participating in community endeavors, while increasing the visibility of local farmers market, and drawing a new audience for the purchase of value added locally grown products.

And Beyond

As a result of a survey indicating farmers were interested in additional value added training in food processing on a commercial level, a Kitchen to Market Workshop was developed. The daylong interactive workshop explored all issues involved in starting a commercial food processing business beyond the farmers market. Twenty-seven farmers and 34 agency professionals from across the stated attended. In addition, a hands-on canning workshop was developed and presented to interested community members. All of the participants indicate they have the necessary knowledge and technical skills to follow safe processing methods using USDA-recommended recipes. Follow-up questions on all training shows behavior changes towards following safe processing methods. As more homebased microprocessors try their hand at value added products, more training will be needed following farm to table concepts. Cooperative Extension professionals are a resource for many farmers who want to start a food business, offering advice at the growing level, workshops on value added products, and guidance on business development.

Other states have begun to rewrite their food codes to allow farmers the sale of jams and jellies made at home using locally produced fruits and vegetables. Tennessee was the first state to ask for a review (in 2004) of their change in legislation that allowed homebased processing of jams and jellies by farmers. In 2005, I visited with the community partners interested in the homebased microprocessing program in Arkansas. They are currently in the consensus building stage. Similar requests have been made from New Mexico and California. It is interesting to note, that all these requests were initiated by farmer groups, lending to the credence that farmers are interested in expanding their market opportunities, and being a viable part of the food system in their community. Along with the Cooperative Extension Service at local universities, they are gathering together their teams to explore homebased microprocessing in their states. This new dimension for local food systems gives local producers an increased capacity beyond just picking and selling and illustrates how unconventional partners can uniquely collaborate to start something new.

REFERENCES

Bastin, S. (2006). Homebased Processor and Microprocessor Brochure. University of Kentucky Cooperative Extension Service. Retrieved October 18, 2007 from http://www.ca.uky.edu/agc/micro.

Brown, A. (2002). Farmers Market Research 1940-2000: An Inventory Review. *American Journal of Alternative Agriculture* (December 2002). 17(4): 167-176.

Ernst, M., & Woods, T. (2006). Horticulture Trends. In Inman, L., Meyer, A.L., & Wilkie, G. (Eds.), *The Kentucky Agricultural Economic Outlook for 2006*. University of Kentucky Cooperative Extension Service.

Dahlberg, K. A. (1999). The Restructuring of Food Systems: Trends, Research and Policy Issues. *Agriculture and Human Values*. 16(2): 109-116.

Feenstra, G. & Campbell, D (1996-97). "Steps for Developing a Sustainable Community Food System," *Pacific Northwest Sustainable Agriculture: Farming for Profit & Stewardship*. 8(4): 1, 6.

Hinrichs, C. C. (2000). Embeddedness and Local Food Systems: Notes on Two Types of Direct Agricultural Markets. *Journal of Rural Studies* 16(3): 299-303..

Kentucky Department of Agriculture. (1996). *Kentucky Farms and Markets: Emerging Policy Opportunities*.

Peters, J. (1997). Community Food Systems: Working Toward a Sustainable Future. *Journal of American Dietetic Association*. 97 (9): 955-956.

Renting, H., Marsden, T. K. & Banks, J. (2003). Understanding Alternative Food Networks: Exploring the Role of Short Food Supply Chains in Rural Development. *Environment and Planning* A 35(3): 393-411.

Susskind, L., McKeaman, S, and Thomas-Larmer, J. (1999). *The Consensus Building Handbook: A*

*Comprehensive Guide to Reaching Agreement***.** Thousand Oaks, CA: Sage Publications.

Thomson, J. S., Radhakrishna, R.B., Maretzki, A.N., and Inciong, L.O. (2006). Strengthening Community Engagement Toward Sustainable Local Food Systems. *Journal of Extension* 44(4); Retreived October 18, 2007 at http://www.joe.org/joe/2006august/a2.shtml

USDA, Agriculture Marketing Service (AMS). (2002). *US Farmers Markets 2000: A Study of Emerging Trends.*

USDA, Agriculture Marketing Service (AMS), Marketing Services Branch. (2004). *Farmers Market Growth.*

USDA, Sustainable Agriculture Research and Education (SARE). (2007). "What is Sustainable Agriculture?" Retreived October 18, 2007 at http://www.sare.org/publications/explore/index.htm

Index

Page numbers in **bold** type refer to **figures**
Page numbers in *italic* type refer to *tables*

agrarian fundamentalism 4, 39, 40, 43, 50–3, 55
Agri-food 39–59; education 54; globalization 48; localization 43; policy development 42
agritourism 92
Amish/Mennonite community 27–38
Appalachia 28
As You Sow (Goldschmidt) 11
Atrazine 11

Barkley, D.: and Henry, M. 11; and Wilson, P. 79
Bastin, S. 5–6
Beaulieu, L.: and Jordan, J. 1–7
benchmarking 30
Benford, R.: and Snow, D. 42–3, 49
Bernat, G.: and Gibbs, R. 10
Bhavsar, V.: and Tanaka, R. 2
biological management 11
Bradshaw, T. 41; and Pigg, K. 40, 45
Brasier, K.: *et al* 3, 5, 8–22
Brundtland Commission (World Commission Environment and Development Report 1987) 1
business 70; development 98

California Alliance for Sustainable Agriculture 74
Campbell, D. 74; and Feenstra, G. 92
capital 63; cultural 75; social 63, 75
capitalism: competitive 43
catalyst farmers 72
Chesapeake Fields (Maryland) 14, 19
civic agriculture 2, 4, 39, 43, 50–3, 55
Clemson University: Vegetable Research Center 81
clusters 3, *12*, 79; benefits 12; characteristics 10; definition 8–9; impact 10–12; infrastructure 17; regional farming 3–4; small farm 8–22
collaboration 40–1; multi-stakeholder 4, 39–59, 55

community 2, 16–21, **72**, 75; agency 4, 6; communication 69; development 24, 28, 34–5, 45, 53, 87; economy 92; networks 63, 75; pride 74; rural 34–5, 75
competition 8–22
competitive capitalism 43
conflict: environmental 41
Conner, D.: Wright, W. and Score, M. 4, 39–59
Cooperband, L.: Gasteyer, S. and Hultine, S. 4
Cornell Cooperative Extension 15–16
County Cooperative Extension Service Horticulture Agent 93
crop management 84
Crosswind Llama and Herb Farm 97
cultural capital 75

dairy graziers 4; Ohio study 4, 23–8; Wisconsin 32
dairy industry 16; confinement *v* grazing 26–7; environmental impact 28; goals 27; health issues 27; pasture-based 23–38; research needs 29
Dave's Supermarket project 63–75, *70*
differentiation 79
diversity 54–5
Donnelley Foundation 81

economic viability 23, 27, 33
economy 1, 33–5; development 71, 92; global 42; growth 9, 20, 34–5; health 91–9; impact 84; infrastructure 33; local 25, 61, 81; profitability 77, 81, 92, 97; rural 62; viability 82
ecosystem management model: grassroots (GREM) 4, 23, 24–5, 34–5
education 20, 67, 95; Agri-food 54; finance 30; health and nutrition 86
emergency food assistance system 77–90
emissions 85
employment 9; natural resources 24
enterprises: value-added 92

environment 1, 51, 77, 79; conflict 41; damage 1, 26; management 11; stewardship 54, 92; vulnerabilities 48
Environmental Programs/Partnerships in Communities (EPIC) 45
equity 1, 53, 83; social 1, 4, 53
extension educators 44

families: low-income 41, 77–90
Family Farm Field Day (Holmes County Amish) 33
Farm Bill (1985) 1
Farm-to-School Project 51
FarmAid tour (2005) 69
farmers: catalyst 72; Hispanic immigrant 14
Farmers' Markets 63–75, 79, 91–9; law 94; Summits 93
farming: large-scale 79; regional clusters 3–4
farms: small 8–22, 27, 34, 61–76, 77–90
Feenstra, G.: and Campbell, D. 92; and Garrett, S. 42
finance 30; education 30; Financial Lending Opportunities in New Agriculture 30; loans 30–1, 71, 83
Finger Lakes Wine Cluster (New York) 11
Flora, C. 63
food 2, 79; assistance 3; banks 5–6; emergency assistance system 77–90; growing locally program (GFL) 5, 78–90; niche 3; safety 86, 94; security 50–5; systems 2, 5, 35, 39–59, 61–76, **66**; transportation 85
food bank: Lowcountry (LFB) 5, 78–90
Food and Drug Administration (FDA) 96; Better Process Control School 96
Food Routes 69
Food System Economic Partnership (FSEP) 4, 39, 44, *46*; goals 45
framing 43, 49
fresh produce enhancement 82
fundamentalism: agrarian 4, 39, 40, 43, 50–3, 55

Gabe, T. 10
Garrett, S.: and Feenstra, G. 42
Gasteyer, S.: Hultine, S. and Cooperband, L. 4
genetically modified (GM) organisms 48
Gibbs, R.: and Bernat, G. 10
global economy 42
globalization: Agri-food 48
Goffman, E. 42
Goldschmidt Hypothesis 79
Goldschmidt, W. 11, 52, 78
Granovetter, M. 40
grassroots ecosystem management model (GREM) 4, 23, 34–5; theory 24–5
graziers: dairy 4, 23–8, 32
grazing: managed intensive (MIG) 29; rotational systems 25
greenhouse gases 85
Growing Foods Locally (GFL) program 5, 78–90
growth: economic 9, 20, 34–5; industrial 18
Guptil, A.: and Wilkins, J. 73

health: dairy industry issues 27; economy 91–9
health and nutrition 80, 83, 92, 95; education 86
Healthy Grown potato 11
Henry, M.: and Barkley, D. 11
Hilchey, D. 12
Hispanic immigrant farmers 14, *see also* New Farmer Development Project (NFD)
Hmong Farmers (Massachusetts) 15, 17
home-based processing 5
Honey Tree Farm 97
horizontal information exchange 25
Horrigan, L.: *et al* 1
House Bill 391 (2003) 5, 91–5
Hultine, S.: Cooperband, L. and Gasteyer, S. 4, 61–76; *et al* 5
Hunt, S.: and Monney, S. 43
Huxham, C. 41

identity: local 19
Ikerd, J. 79
Illinois: Central 5, *see also* food systems; Local Food Project
immigrant farmers: Hispanic 14
IMPLAN input-output model building procedure 80, 85
independent retail 73
industrial growth 18
infrastructure 27; development 53; economic 33; social 31–2, 61
innovation 8–22, 41
inventory tracking system 86
Ireland 28

Jordan, J.: and Beaulieu, L. 1–7

Kentucky 5–6; Cabinet for Health Services 91; Department of Agriculture 91; Farm Bureau Certified Roadside Markets 91; Farmers Market Association 93; *Farmers Market Manual* 93; law 94; University Cooperative Extension 91
Kettl, D. 41
Killbuck Creek 32
Kitchen to Market Workshop 98

land use 47

large-scale farming 79
leadership 73, 78; local 3
Lee, J.: *et al* 11
legislation 5, 91, 94, 98
Lenawee Ethanol Production Facility 49
loans 30–1, 71, 83
local economy 25, 61, 81
Local Food Project (Fairbury, Illinois) 61–76, **64**, *65*, **66**, *72*
local identity 19
local leadership 3
localization: Agri-food 43
low-income families 41, 77–90
Low-Input Sustainable Agriculture program 1
Lowcountry Food Bank (LFB) 5, 78–90; mission 86
Lowcountry (South Carolina) 77–90
Lowndes, V.: and Skelcher, C. 41
Luloff, A.: and Swanson, L. 6
Lyson, T. 2, 52, 63

McAdam, D. 49
McCall, T. 42
managed intensive grazing (MIG) 29
management 11, 26, 77, 84
marketing 61; direct 66–75, 91, 93
Mennonite community 27–38
mental well-being 15
Meter, K. 62, 73
Michigan: Coalition of Black Farmers 44; Farmers Union 44; Organic Food and Farming Alliance 44
microbiology 96
microprocessing 91–9, **96**; homebased 91, 95–7
Monney, S.: and Hunt, S. 43
Morton, L.: and Miller, L. 4, 23–38
multi-stakeholder collaboration 4, 39–59, 55

natural resources 24, 77
networks: community 63, 75
New Agriculture 30
New Farmer Development Project (NFD) 14
New York Certified Organic (NYCO) 13–14, 17, 19
New York City Greenmarkets 14
New Zealand 28
North Central Ohio Dairy Grazing Council (NCODGC) 32
North Country Dairy Viability Initiative (NCDVI) 15, 17, 19

Ohio Agricultural Research & Development Center (OARDC) 27, 34; Agroecosystem Management Program 29
on-farm strategies 2
Organic Crop Improvement Association (OCIA) 13
Organic Growers of Michigan 44
organic produce 64, 76

pasture: management 26
Pennsylvania: Tuscarora Organic Growers (TOG) 16; Women's Agricultural Network (PA-WAgN) 15, 17
Peters, J. 92
Pigg, K.: and Bradshaw, T. 40, 45
Piore, M.: and Sibel, C. 79
Pirog, R. 85
Porter model 79
pre-industrial/chemical-petro farming era 25
producer frames 43
products: value-added 5, *93*
profitability 2; economic 77, 81, 92, 97
Public Act 207 (1921) 47
public relations 69, 71

quality of life 2, 11, 28, 42, 54, 78, 92

regeneration 41
regional development 8–22, 44
resources: natural 24, 77
retail: independent 73
Robinson, K.: *et al* 5, 77–90
rotational grazing systems 25
Rubbermaid 32
rural community 34–5, 75
rural development 23–38, 53
rural economy 62
rural-urban interface zones 26

Score, M.: Conner, D. and Wright, W. 4, 39–59
self-reliance 80, 82
Sibel, C.: and Piore, M. 79
Silicon Valley 9
Skelcher, C.: and Lowndes, V. 41
Slow Food Huron Valley 44
Small Farm Institute 27, 34
small farms 8–22, 61–76, 77–90
Smithers, J. 62, 75
Snow, D.: and Benford, R. 42–3, 49; *et al* 42
social capital 63, 75
social equity 1, 4, 53
social infrastructure 31–2, 61
social viability 23
soil management 84
South Carolina 5, *see also* food banks
Southern Rural Development Center 2
stewardship: environmental 54, 92
Sustainable Agriculture Research and Education (SARE) 1

Sustainable Community Innovations Grant Program (SCIG) 2–3
Swanson, L.: and Luloff, A. 6

Tanaka, R.: and Bhavsar, V. 2
technology: development 10, 23, 25
tobacco 91
top down/blueprint planning 39
tourism 4, 11, 18, 71, 81, 92
training 9–10, 41, 86
transport 85
Tuscarora Organic Growers (TOG) (Pennsylvania) 16

United Kingdom (UK) 28
United States Department of Agriculture (USDA) 27; Agricultural Research Service (ARS) Coshocton Branch 29; Fund for Rural America 41; National Organic Program 68; Seasonal Dairy Grazing Project 29; Sustainable Agriculture Research and Education 92

Value Added Microprocessor Project 96
Vegetable Research Center (Clemson University) 81
viability 21, 66, 82; economic 23, 27, 33; social 23

wages 9
water management 84
wildlife 26
Wilkinson, K. 53
Winne, M.: *et al* 41
Women: Infants and Children (WIC) 93
Woodford County Twilight Festival 93
World Commission Environment and Development Report (Brundtland Commission 1987) 1
Wright, W.: *et al* 5; Score, M. and Conner, D. 4, 39–59